# SACRED EVOLUTION

## Bridging Science and Faith
## as Partners for the Future of Humankind

Hasan Gokal MD

**Sacred Evolution**
*Bridging Science and Faith as Partners for the Future of Humankind*

**Published by:**

**Hasan Gokal MD**

www.gokalmd.com

**ISBN Information:**

Paperback: ISBN 979-8-9991154-0-9

Hardcover: ISBN 979-8-9991154-2-3

E-book: ISBN 979-8-9991154-1-6

Library of Congress Control Number: 2025911499

**First Edition:** June 2025

**Printed in the United States of America**

**Disclaimer:**
The views expressed in this book represent those of the author. While every precaution has been taken in preparation of this book, the author assumes no responsibility for errors, omissions, or damages resulting from the use of the information herein.

For permissions, inquiries, or other requests, please contact:
**Hasan Gokal, MD**
hasangokal@gokalmd.com

# Acknowledgments

Writing *Sacred Evolution* has been a sacred journey, one blessed by the generosity, wisdom, and love of many remarkable people.

My deepest gratitude belongs first to my beloved wife, Maria Gokal. Her unwavering support, encouragement, and spiritual strength have sustained me throughout this journey. Maria's thoughtful proofreading and insightful reflections infused the very heart of this book. I am endlessly thankful for her partnership, patience, and faith.

To our wonderful children, Zahra, Ali, and Mariam, I offer heartfelt thanks. Each of you supported me in your own unique and loving ways, filling our home with warmth, laughter, and encouragement. Your belief in this project and in me has been an immense source of inspiration.

My profound appreciation also goes to my parents, Kassim and Nasim Gokal, whose steadfast values and quiet wisdom provided the core inspiration for *Sacred Evolution*. The principles you exemplified shaped this work profoundly, and I hope it stands as a tribute to the beautiful foundation you provided me.

Special thanks go to Dr. Hasnain Walji, S. Farhat Abbas, and Dr. S. Moustafa Qazwini for their invaluable religious guidance, careful fact checking, and consistent support. I am grateful also to Emily Dunn, and Chris Block, whose encouragement and scriptural guidance greatly strengthened the integrity of this book.

I owe special gratitude to my dear friend, David Oates, whose encouragement, moral support, and wisdom in reaching the right audience guided my steps. And to Snighda Nandipati, a talented published author whose enthusiastic encouragement and expert guidance illuminated my path to publication, I thank you sincerely. Finally, my heartfelt appreciation extends to everyone who contributed, encouraged, and supported this book's creation, both directly and indirectly. Your kindness has left an indelible mark, and I remain deeply humbled by your generosity and companionship on this sacred journey.

# Preface

This book explores the profound relationship between spiritual traditions and scientific understanding, highlighting how ethical virtues from religious texts guide humanity towards a cooperative and resilient future. Intended for thoughtful readers interested in **bridging science and spirituality**, it aims to foster dialogue, understanding, and mutual respect across different faith traditions and secular perspectives.

## On the Inclusion of Religious Traditions

While this book draws extensively on Islamic and Christian sources to illustrate the Evolutionary-Spiritual framework, it is important to emphasize that this approach is not exclusive to these two traditions. The same evolutionary patterns, emphasizing cooperation, moral development, inner discipline, and collective responsibility, are evident in many other religious and philosophical traditions as well. Foundational texts and teachings from **Judaism, Hinduism, Buddhism, Sikhism, and even secular humanism** often convey principles that align remarkably well with the moral and cooperative trajectory discussed throughout this work.

Future volumes may explore these traditions more fully, incorporating their parables, ethical prescriptions, and spiritual practices into the larger vision of humanity's cooperative and evolutionary future. The present focus on Islam and Christianity is simply a matter of **scope and familiarity**, not preference or exclusivity.

## On the Use of AI Assistance

This manuscript was authored with the assistance of artificial intelligence tools for the purpose of **language refinement, editing support, and structural organization**. However, the core ideas, conceptual frameworks, philosophical positions, interpretive insights, and overall vision presented throughout the book are entirely my own as the author.

AI was used to articulate and clarify what was already **conceived and structured through deep reflection, interdisciplinary reading, and years of personal inquiry**. It is not an AI-generated book; it is **a human-authored work aided by technology**. The intention behind this disclosure is to ensure transparency while affirming the authenticity and **originality of the content.**

# Introduction – Bridging Science and Religion

Historically, **religion and science have often been at odds**. Scientific discoveries in fields such as physics, chemistry, and biology have brought clarity to phenomena once attributed exclusively to divine or supernatural explanations. A functional understanding of these scientific fields is indispensable today. Therefore, it is unsurprising that younger generations, deeply educated in scientific methods and empirical reasoning, often perceive religious teachings as **antiquated and disconnected** from reality.

This perception has led to disillusionment, particularly among those who feel compelled to abandon religious traditions that appear **incongruent with their understanding of the natural world**. Stories from sacred texts (such as those in the Bible or the Quran) can seem allegorical or mythological, causing individuals to feel conflicted or even dishonest when trying to uphold traditional beliefs. In previous eras, unexplained phenomena such as eclipses, lightning, earthquakes, and floods **actually strengthened religious faith** precisely because they required faith in the unknown. Today, events like eclipses are fully explained by astronomy (the moon blocking sunlight), eliminating the mystical component once needed to bridge gaps in human understanding.

To further complicate matters, these issues are compounded by some well-meaning religious advocates without science backgrounds who use explanations and analogies that fail to resonate with educated, empirically minded young people. These **further distances the youth from appreciating spirituality and religion**. Additionally, cultural traditions and recurring disputes among religious authorities, such as disagreements over moon-sighting in Islam or theological differences within Christian and Jewish traditions, exacerbate these feelings of alienation.

This book seeks **to re-frame the dialogue between religion and science** in a way that eliminates this ever-growing gap. But it is much more than that. It forces one to step outside the fray and **re-introduces religion and science as partners towards a greater goal**. It brings together the teachings of the great religions of the world and **introduces a perspective which is impossible to ignore**. Everything should start to make sense in a way that never did before. Upon completing this book, the reader should be able to recognize the bridge over the perceived gap between religion and science through an approach that integrates empirical observation with spiritual insight. With each section, examples are presented primarily from **Christian and Islamic** sources. The limitation to Christian and Islamic sources is for no reason other than brevity. When comparable lessons, and parables are compared from other texts (Judaism, Hinduism, Buddhist, and even Secular Humanist), the **same pattern of cohesiveness and unity emerges**. The fundamental principles apply, and it is hard to ignore the unifying direction that is evident in this book. The objective is not to provide exhaustive explanations or authoritative guarantees about every scientific or religious claim, but rather to offer a **coherent set of observations and encourage readers to integrate them** meaningfully within their personal belief systems.

## The progression of this book is structured as follows:

First, it **explores scientific knowledge** from the birth of the universe (the Big Bang) onward. Detailed quantum physics is intentionally excluded to maintain relevance and readability for most readers. It describes the spontaneous formation of simple atomic particles, leading progressively to more complex molecules.

Subsequently, the narrative **transitions to biological evolution**. It progresses from the formation of simple cells to complex multicellular organisms and eventually reaches humans. Importantly, the text proposes that human evolution is likely ongoing, speculating that **humanity will continue evolving into even more complex, integrated forms** potentially unrecognizable by today's standards.

Next, the discussion moves to the essential directions humanity must follow to evolve toward these more advanced states. Here, **religious teachings are introduced as crucial guidance** to help humanity reach these higher developmental stages. Examples demonstrate how **religious teachings align closely with this scientific and philosophical approach**, showing consistency across multiple scenarios.

Through this integrative perspective, readers deeply rooted in scientific traditions may find a renewed appreciation for and understanding of religious concepts, **reconciling discrepancies** that have previously fostered alienation. The author, as both a scientist and physician who has personally navigated this reconciliation, offers these observations to aid others in bridging their own gaps between empirical science                          and                          spiritual                          belief.

# Sacred Evolution

# Part I: Foundations

# Chapter 1: A Quick Summary of the Evolutionary-Spiritual Bridge Framework

The central premise of this work is that **humanity's current state represents merely an intermediate step within a much broader evolutionary trajectory**. Human beings have progressively evolved physically, psychologically, intellectually, and morally, from simpler cosmological origins to their present form. However, this process is far from complete.

Drawing from principles in **thermodynamics**, evolutionary **biology**, **philosophy**, and deep ethical traditions found in religious teachings (particularly **Biblical** and **Quranic** traditions), this approach suggests that humanity continues to be propelled forward along a path of increasing complexity, interconnectedness, ethical sophistication, and cooperative intelligence.

Just as the **universe evolved spontaneously from simplicity to complexity**, beginning with basic atomic interactions and culminating in intricate biological systems, human societies similarly evolved from **basic survival-based** interactions **to sophisticated ethical and cooperative systems**. From examining the past, it would be reasonable to presume that **humans may evolve further, into yet unrecognizable forms.** Going forward, humans of today **may not even be able to recognize the greater organism** we will comprise, just as a cell would not recognize the larger organism it is part of.

This is where **religion enters the discussion**. Based on this view, the moral and spiritual teachings found in most **religious traditions can be seen as evolutionary nudges** that encourage humanity to develop in ways that **ensure long-term survival** and possible development to ever

**greater capabilities**. In essence, these teachings serve as purposeful evolutionary nudges: **deliberate ethical prescriptions** designed to facilitate humanity's growth towards a **state of advanced collective intelligence.**

In this evolutionary-spiritual perspective, concepts such as **love, justice, mercy, patience, education, stewardship of the environment**, and **compassionate care** for vulnerable populations are not just ethical virtues; they are evolutionary necessities. Conversely, behaviors like **pride, greed, envy, substance abuse, exploitation**, and **violence** explicitly **represent evolutionary barriers**, limiting humanity's collective advancement and increasing social, moral, and ecological entropy.

Ultimately, **this approach offers a cohesive, integrative vision:** future human evolution is understood as a **purposeful trajectory** toward greater ethical cooperation, social harmony, intellectual sophistication, emotional resilience, ecological responsibility, and spiritual insight; characteristics critical for **ensuring humanity's sustainable future and continued evolutionary success.**

In what follows, we will first briefly explore the scientific foundations underlying this perspective, beginning with **cosmological origins, chemical evolution, and biological emergence**, before moving toward human societal development and its associated complexity. Don't let the scientific discussion intimidate you. This book is written so that the core ideas will remain clear and meaningful even if you choose to skim or skip some of the scientific details.

After this initial overview, we will delve deeply into the **ethical, philosophical, and spiritual dimensions** that form the heart and essence of the evolutionary-spiritual journey. This intellectual approach integrates empirical scientific understanding with principles of cooperation and synergy, offering **a coherent pathway toward future human sustainability**. Recognizing cooperation as both a scientifically validated strategy and a principle echoed in spiritual teachings positions humanity to address imminent resource and environmental challenges effectively. Perhaps within this **profound alignment of science and spirituality** lies the true understanding of what many traditions call the divine: a force driving the cosmos and life toward greater unity and complexity.

# Chapter 2: The Science – Evolution from Cosmological Origins to Human Behavior

## 1. Cosmological Origins and Particle Physics

The narrative begins with the **Big Bang**, a cosmological event characterized by intense energy, quantum fluctuations, and the **formation of subatomic particles**. Before the first parts of the atom (Neutrons, Protons and Electrons) came into being, the precursors came into existence. Early cosmic conditions allowed particles such as **quarks and leptons** to stabilize into **protons**, **neutrons**, and **electrons**, eventually **forming hydrogen atoms**. Through gravitational forces, these hydrogen clouds **accumulated into stars**, initiating nuclear fusion that **created heavier elements** (like carbon, oxygen, and nitrogen) essential for life. In this stage, the universe transforms from an extremely hot, dense state to a cooling and expanding cosmos filled with the building blocks of matter, governed by **thermodynamic principles**. The key point in this section is to appreciate that after the initial release of this energy at the point of the origin, each new building block occurs **spontaneously**, without constant moment by moment direction from any source. Each new substance that comes into being, does so **guided by the rules of energy conservation** (path of least resistance). Thinking about this is sometimes difficult, until one recognizes that this occurs constantly around us unnoticed. For example, combining carbon with oxygen to release a constant stream of carbon dioxide from burning wood, requires no more intervention, than quarks coming together to form protons, or protons and electrons coming together to form hydrogen.

## 2. Chemical Evolution and Molecular Complexity

When **stars run out of hydrogen**, they start **forming heavier elements**. It makes sense that 2 hydrogen atoms (1 proton each) when fused together, will likely form a Helium atom (2 protons each). Heavier atoms (with more protons and neutrons), combine with other atoms to form heavier still atoms. Eventually, **a star runs out of fuel**, gets bloated, expands, and eventually explodes. This explosion of a u**sed-up star is called a supernova**. Supernova events disperse heavy elements into space, many of which clump together under gravity and **form planets**. Chemistry took shape: simple molecules ($H_2O$, $CH_4$, $NH_3$, etc.) formed in these new environments of young planets. On early Earth and likely other planets, conditions allowed complex molecules to emerge. For example, **carbon's unique bonding capabilities** led to an array of **organic compounds**. Through processes not yet fully understood (but often demonstrated in laboratory simulations), these **molecules self-organized into structures like amino acids and nucleotides**, governed by **chemical thermodynamics** and **reaction kinetics**. Molecular self-assembly processes yielded increasingly complex organic molecules, including **amino acids**, **nucleic acids**, and **phospholipids**, foundational components for **prebiotic** chemistry.

### 3. From Chemical Self-Assembly to Self-Replication: Thermodynamic Underpinnings

A critical scientific challenge lies in connecting simple chemical self-assembly to **self-replicating molecular systems**. Complex chemical networks eventually gave rise to self-replicating molecules. The details of how this happens, includes various mechanisms, including autocatalytic sets which are systems of molecules that catalyze each other's formation, thus setting up a simple form of reproduction. **Energy flows** in accordance with the second law of thermodynamics ultimately drive these mechanisms.

For those inclined towards a more technical definition, spontaneous assembly and replication processes occur when the Gibbs free energy of the system decreases ($\Delta G < 0$), increasing stability and reducing energy states. Early molecular environments, rich in chemical gradients and energy fluxes (e.g., geothermal vents, UV radiation), facilitated energetically favorable polymerization reactions. **In simpler terms**, early raw materials of life harnessed energy to build order (e.g., replicating RNA strands) in a way that was favored by the environment. Through repetitive cycles of replication, selection, and mutation, driven solely by thermodynamic principles and environmental energy gradients, simple self-assembling systems **evolved into stable self-replicating entities**. This stage marked the transition from chemistry to biology: when molecules began to copy themselves and evolve by a process reminiscent of natural selection.

The complexity and details of these mechanisms are far beyond the scope of this book. As such, the details of these mechanisms are left out from this book but are easily available to those inclined to study them further.

## 4. Biological Evolution and Increasing Complexity

Pre-biologic chemical evolution gradually transitioned into biological evolution with the **emergence of self-replicating RNA and DNA molecules**, which introduced genetic inheritance and natural selection into the process. Simultaneously, **lipid bilayers formed primitive cellular membranes**, creating isolated bubbles of micro-environments conducive to biochemical complexity. As early cellular life evolved, sophisticated molecular machinery developed to efficiently manage energy usage, store and transmit genetic information, and enable adaptive responses.

The **critical evolutionary transition** from single-celled organisms (prokaryotes) to complex multicellular life (eukaryotes) was driven by environmental pressures, symbiotic relationships, and genetic diversification. **Endosymbiosis** then occurred, which is when single-celled organisms integrated to form cells containing specialized structures such as nuclei and mitochondria. This marked a significant milestone, enabling the conditions for further biological advancement.

**Multicellularity** allowed cells to specialize, communicate, and cooperate as unified organisms, paving the way for life's remarkable diversification. Over geological time, this progression led from simple multicellular organisms to the **Cambrian explosion**, a dramatic **expansion in biological complexity**, eventually yielding diverse lineages such as fish, amphibians, reptiles, mammals, and ultimately, primates.

Each evolutionary step exemplified increasing biological integration: individual cells forming colonies, colonies developing into complex organisms, and organisms cooperating within sophisticated social groups, demonstrating the profound power of evolutionary cooperation and complexity.

## 5. The Emergence of Human Intelligence and Society

Humans represent a recent development in this long evolutionary story. Homo sapiens evolved a large brain **capable of abstract thought, language, and foresight**. This allowed rapid cultural evolution on top of biological evolution. Humans formed complex societies with social structures, technologies, and eventually civilizations. Initially manifesting as individual survival strategies, **evolutionary benefits of cooperative behavior became pronounced**, leading to complex social structures.

Crucially, humans developed **ethical** and **spiritual** frameworks, otherwise known as cooperative codes of behavior, that extended our social cohesion beyond genetic or familial ties. Some view this as a continuation of evolution: just as multicellular organisms require cooperation among cells, **advanced societies require cooperation among individuals**.

This cooperative approach may already be deeply embedded within human biology and psychology. Phenomena such as burnout, dissatisfaction, and existential dread often arise from a perceived lack of meaningful contribution to society. Ernest Becker's "Denial of Death" introduces the concept of "**immortality projects**," where humans seek enduring legacies to achieve psychological contentment and a **sense of meaningful existence**.

Modern examples support this view. The success of Wikipedia, developed collaboratively by millions without financial incentive, surpassing traditional commercial encyclopedias, illustrates the innate human drive to contribute meaningfully and **"leave a mark."** This becomes important later on because this behavior aligns seamlessly with Islamic and biblical principles advocating for collective good, social responsibility, and lasting contributions to society.

Throughout these scientific stages, one can observe a **trajectory toward greater complexity and integration**. From particles coming together to form atoms, atoms to molecules, molecules to living cells, cells to organisms, and organisms to societies, **the arc of evolution** (cosmic, chemical, and biological) trends toward **higher levels of organized complexity**. Sociobiological theories, including, reciprocal altruism, and group selection, explain the evolutionary advantages of cooperative behaviors in species ranging from insects to mammals, and including humans.

## 6. Epigenetics and Behavioral Inheritance

While much of what we have discussed so far involves evolution through genetic changes over vast timeframes, another more immediate and flexible mechanism is now emerging, one that may link our behaviors, traumas, and moral decisions directly to the biology of our descendants. This mechanism is called epigenetics.

As we trace humanity's evolutionary journey from particles to societies, one surprising and relatively recent scientific insight stands out: epigenetics, the study of how **life experiences can shape gene expression and potentially be passed to future generations**. This concept challenges the older idea that genetic inheritance is purely fixed and unaffected by the environment. Instead, epigenetics reveals a powerful and dynamic relationship between what we experience in our lifetime and how **our descendants may be biologically influenced** by it.

*What Is Epigenetics?*

At the most basic level, every cell in the human body contains the same set of DNA, a genetic "blueprint." But **not all genes are turned on** or off in every cell. For example, the genes that make a heart cell function are different from those active in a brain cell. **Epigenetics refers to the molecular markers that determine which genes are expressed** (activated) or silenced (deactivated), depending on internal signals and environmental influences.

These changes **do not alter the DNA code itself**. Instead, they involve chemical tags, such as methyl groups, that attach to the DNA or to the proteins that package it, called histones. These tags act like **switches or dimmers**, telling genes whether to be loud, quiet, or completely silent.

What is truly remarkable is that some of these epigenetic changes can be passed down to children and even grandchildren. That means your diet, stress levels, trauma, exposure to toxins, and even emotional states could leave a **biological imprint** not only on you but also on your descendants.

## Examples of Epigenetic Influence

1. Starvation and Resilience

One of the most famous studies on epigenetics comes from the **Dutch Hunger Winter** of 1944 to 1945. During a severe famine, pregnant women in the Netherlands experienced extreme malnutrition. Decades later, scientists discovered that the children born during that period had higher rates of obesity, diabetes, and heart disease, even though they were raised in more plentiful times. These children's genes had been chemically marked by their mothers 'starvation, preparing their bodies to conserve energy and store fat. This trait made sense during famine but became maladaptive in a post-war world.

## 2. War and Psychological Trauma

Similar patterns have been found in populations exposed to extreme psychological trauma. Studies of **Holocaust survivors** and their children suggest that the biological stress responses, such as heightened cortisol sensitivity, may be epigenetically transferred. Trauma, in this view, is not only psychological; it leaves a **physiological signature on the body** that may affect emotional regulation and stress resilience for future generations.

## 3. Paternal Effects and Lifestyle

While maternal influence is more intuitively understood, since the womb environment is critical, **fathers 'experiences matter** too. Research has shown that men who began smoking before puberty were more likely to have sons with higher body fat. Other studies suggest that a father's stress or diet prior to conception can influence the sperm's epigenetic profile, which may in turn alter the developmental trajectory of the offspring.

### *What Does This Mean for Human Evolution?*

Traditionally, evolution has been described as slow genetic change driven by random mutation and natural selection. Epigenetics adds a faster and more flexible layer of adaptation. It means that **human beings are not only shaped by their genes but also by what they do, how they live, and what they suffer or overcome, with potential consequences rippling through generations**.

This insight has profound implications for the evolutionary-spiritual model. It suggests that behavioral patterns, spiritual disciplines, communal traumas, and social environments do not merely vanish with death. Instead, **they may become part of our biological legacy**.

*Consider this:*

A generation that grows up amid conflict and fear may pass on epigenetic markers associated with anxiety and hypervigilance.

A community that embraces mindfulness, spiritual calm, and ethical restraint may biologically prime the next generation for resilience and balanced neurochemical regulation.

Chronic inequality, stress, or systemic violence may embed biological disadvantages across generations unless these conditions are addressed not only through social change but also through healing and moral reform.

## The Ethical and Spiritual Dimension

Epigenetics thus adds a new layer of urgency and meaning to how we live. Our moral choices, social systems, and spiritual practices are not just momentary; **they are inherited experiences, encoded biologically**. This affirms a deeply spiritual idea: we are the stewards not only of our Earth but also of our descendants 'inner landscapes.

From an evolutionary perspective, the rise of epigenetics suggests that humanity's trajectory is increasingly shaped by **conscience and behavior** rather than by chance mutation. As we choose cooperation over conflict, reflection over impulse, and compassion over domination, we are not only bettering ourselves but also potentially reshaping the human genome in ways that make goodness heritable.

In this light, spiritual prescriptions, such as the ethical teachings found in religious traditions, take on new power. They are not merely rules from the past; they may function as **evolutionary tools** designed to shape human beings biologically, socially, and spiritually, and to create a future that is not only sustainable but also meaningfully integrated.

As we will now explore, human society itself may be evolving toward a form of distributed intelligence in which the collective becomes more than the sum of its parts.

**Current Scientific Debates and Consensus on Epigenetic Inheritance**

The concept of epigenetic inheritance has sparked significant excitement and debate within scientific communities. While studies such as those of the Dutch Hunger Winter provide compelling evidence supporting transgenerational epigenetic effects, broader scientific consensus remains nuanced.

Current debates primarily revolve around the mechanisms, extent, and persistence of epigenetic inheritance across generations:

- **Mechanism Clarity**: While DNA methylation and histone modification clearly influence gene expression, how reliably these marks are transmitted through multiple generations remains debated.

- **Replication and Consistency**: Some epigenetic inheritance effects in mammals have proven challenging to replicate consistently across laboratories, highlighting potential variability due to subtle environmental differences and methodological approaches.

- **Duration of Effects**: Short-term inheritance (from parent to offspring) is well-supported; however, long-term, multi-generational inheritance in humans remains less clearly demonstrated.

Current consensus acknowledges epigenetic inheritance as a genuine phenomenon, particularly robust in plants and specific animal models (such as mice). Still, long-term transgenerational epigenetic inheritance in humans requires cautious interpretation and rigorous further research. Recognizing these nuances ensures we approach epigenetics with clarity and humility, underscoring the dynamic interaction between environment and biology without prematurely attributing sweeping hereditary outcomes to all environmental exposures.

### 7. Human Society and Future Evolutionary Trajectories

Human societies embody a sophisticated level of sociobiological complexity, shaped profoundly by **intellectual capacity and collective behaviors**. Today, human evolution faces critical challenges, particularly in the areas of sustainable resource management and environmental adaptation. To navigate these pressures successfully, societies must adopt **elevated forms of collaboration** that reflect biological patterns observed in highly cooperative species.

If humans represent an intermediary step in evolution on a universal timescale, **what might the next steps look like?** Consider this: every act of collaboration between living units begins with some form of communication. In simple cells, communication occurs through **physical** or **chemical** means. Among groups of cells or small organisms, communication expands to include signals conveyed through **light**, **sound**, **electrical impulses**, **chemical exchanges**, or **tactile interaction**. As organisms increase in complexity and intelligence, communication becomes correspondingly **diverse** and **creative**.

In this context, the global emergence of real-time information exchange systems may signify the early formation of a tightly integrated, **collective human consciousness**. This development might be so subtle or dispersed that individual humans are not fully aware of it. Although such an interpretation of human development may seem ambitious, it is not beyond the realm of possibility. It represents a plausible trajectory toward **deeper psychic and social integration across humanity**.

If this idea seems far-fetched, consider the following example. It is commonly assumed that the brain, as the topmost structure of the nervous system, houses the vast majority of the body's neurons. However, the human gastrointestinal (**GI**) tract contains a surprisingly extensive network of nerve cells, numbering approximately **100 million neurons**. This system, known as the **Enteric Nervous System (ENS),** is embedded within the walls of the GI tract and operates semi-independently to regulate digestion while remaining intricately connected to the brain through what scientists call the **gut-brain axis**.

This gut-brain connection has deep evolutionary roots, underscoring the essential coordination between digestive and neurological functions. The gut is also a major site of **neurotransmitter production**. In fact, around 90 percent of the body's **serotonin**, a chemical crucial for mood, sleep, and appetite regulation, is produced in the gut. Other neurotransmitters, such as **dopamine**, **gamma-aminobutyric acid (GABA),** and **acetylcholine**, are also produced or modulated by intestinal cells and the microbiota residing there. These microbes influence not only digestion but also emotional well-being and cognitive functions, including memory, learning, and mood regulation.

Moreover, the GI tract contains trillions of microorganisms whose metabolic byproducts support **brain plasticity**, **reduce systemic inflammation**, and **enhance cognitive health**. The GI tract is not alone in this regard. The **skin** also hosts a diverse and dynamic microbiome that interacts with the human immune system and impacts overall health in ways that researchers are only beginning to understand. These examples reveal previously unrecognized lines of communication between distant systems within the body as well as between the body and the microscopic life forms that inhabit it.

Looking at these unexpected relationships between living entities, which we are just starting to uncover and understand, the above examples demonstrate the many possibilities of how future humans may contribute to a greater collective organism, whether or not they recognize their role in the greater development.

## 8. Introducing Spirituality as a guide for Evolution

So far, we have seen that scientific observations have produced some **repetitive patterns**. The idea of "survival of the fittest" not only describes biologic systems, but a form of it can be applied to pre-biology chemistry as well. For example, when elements are formed, there is more than one configuration for that element to exist in. Some of those configurations are more stable than others. Therefore, most of the type of element will be of the most stable type, and there may be some less stable forms along with it. When you look at the periodic table of elements, and you look at the **atomic mass**, you see that the numbers are not whole numbers as would be expected if you simply counted the mass of the protons and neutrons (remember that electrons have no mass). That lack of exactness comes from the averaging presence of small numbers of elements that have a different number of neutrons, than most of the elements. The average mass is less than or greater than the most common type of element and listed on the periodic table.

The same can be applied to molecules forming. 2 elements can form multiple configurations of the same molecule and are known as **isomers**. Some are more **stable** than the other and constitute much of the molecule type.

Among biologic systems, these natural selection processes are easy to see. We see them among bacteria causing **infections**. When we apply **antibiotics** to treat those infections, the bacteria which are more resistant to the effects of the antibiotics survive, and their offspring all have the ability to resist the antibiotic, causing **antibiotic resistance**.

Among larger animals, this type of natural selection was seen in Africa within the last 50 years. During the Mozambican Civil War (1977–1992), both sides financed their efforts through **ivory poaching**, resulting in the death of approximately 90% of the elephant population in

Gorongosa National Park. This intense selective pressure **favored elephants without tusks**, as they were less likely to be targeted by poachers. Consequently, the proportion of tuskless female elephants increased dramatically, from about 18.5% before the war, to over 50% by the early 2000s.

Among humans, natural selection may take novel forms. An example of this is the ability to understand technology, science, complex engineering systems and other talents may provide certain types of survival advantages to humans, which may be passed onto their offspring.

Evolution is rarely simple and linear and has often been described as taking a meandering path. Much like a river which changes course, based on the changing landscape. You only have look at the earth from an altitude to see many c-shaped lakes formed by the changing course of a river. If we think of evolution similarly, up until now, it was affected primarily by environmental pressures. With the development of **behavior and intellect**, the pressures leading to future evolution does not have to be restricted to environmental pressures. **In a matter of speaking for the first time, we may have a say in what our future evolution will look like!**

The **Evolutionary-Spiritual Vision or Bridge** suggests that teachings in the form of religious and spiritual prescriptions represent a new type of evolutionary pressure, which could not have been possible without both **intellectual and behavioral developments** that humans now possess. **The idea of growing together as a species through the adoption of collaborative behavior is what this new vision proposes**. And when we look back at existing religious teachings, we start to see a **pattern** where the same behaviors are being impressed upon humans, presumably for their benefit. The Evolutionary-Spiritual framework proposes that the benefit of these prescriptions is not just for humans as individuals, but for the evolutionary growth and long-term survival into **more capable organisms** and possibly even as a **collective organism** with a **collective consciousness.**

While current challenges facing humanity, such as climate change, warfare, and resource scarcity, are clearly identifiable, this approach is equally positioned to address unknown future challenges. Fundamentally, many of humanity's historical and contemporary challenges arise from **fluctuations in resource distribution and availability**, inherently driven by thermodynamic principles. Even hypothetical scenarios, such as **cataclysmic events**, ultimately reflect **resource redistribution** and **thermodynamic shifts**. Thus, adopting cooperative, resource-conscious strategies today inherently equip humanity to respond effectively to unforeseen future challenges.

This scientific journey, from the birth of the universe to the rise of human societies, provides an empirical narrative of **increasing complexity** and **interdependence**. Notably, cooperation, whether among subatomic particles forming atoms, organelles within a cell, or individuals within a society, emerges as a recurring theme. Science shows us that **unity and collaborative structures often lead to more stable and complex systems.**

# Chapter 3: Testing Key Biblical Concepts

This Evolutionary-Spiritual framework suggests that **since humans now have intelligence and behavior, we are increasingly in a position to take the reins of our own evolutionary future**. Human can vary between a highly thoughtful, well-balanced existence, to one of carelessness, wastefulness and lack of consideration of others and the environment. **One has the potential to be constructive or destructive to the future of humanity.**

When observing many religious stories, lessons and prescriptions, regardless of the background, **another pattern emerges**. It is increasingly obvious that the stories and the lessons all point towards methods of **guiding human development** towards robust, unified and more intelligent beings. This pattern of thriving from unification is not new. **Bees** survive in colonies, and **wolf packs** thrive when compared to lone wolves. There is an undeniable push from most spiritual teachings to influence human behavior **towards a unified organismal approach**, and **away from individualistic behaviors** which are detrimental to long term survival of humanity and its future forms.

The following section takes some key **biblical** and **qur'anic** teachings and places it against the proposed Evolutionary-Spiritual framework of understanding. The purpose here is to see if the underlying lessons and prescriptions are promoting or destructive towards future evolutionary growth of humans.

These are **far from complete**, and do not include other religion or secular teachings to as to keep the focus manageable within this book. It does not imply a preference towards any one religion, and the reader is encouraged to **apply other spiritual teachings** to see if the framework is universally applicable.

# Key Biblical Stories

## Creation and Stewardship (Genesis 1–2)

The Bible begins with humans created in God's image and instructed to "fill the earth and subdue it" and to be stewards of creation. This implies a **responsibility to care for the environment and all life**, aligning with the idea that humans are meant to cultivate and manage Earth in a cooperative, sustaining manner rather than exploit it destructively. Stewardship suggests an **evolutionary nudge toward living in harmony with nature** (a trait necessary for long-term survival).

## Noah's Ark – Collective Survival through Cooperation (Genesis 6–9)

In this story, Noah cooperates with family and a collection of animal species to survive a global flood. Regardless of one's belief in the literal historicity of the flood, **the narrative exemplifies successful collaborative behavior in the face of catastrophe**. Noah's obedience and teamwork ensured the continuity of life. This highlights how cooperation and collective action are keys to human (and ecological) survival and resilience in times of dramatic environmental challenge.

## Tower of Babel – The Perils of Hubris and Disunity (Genesis 11)

Humanity, speaking one language, attempts to build a great tower to reach heaven. God confuses their language, causing the project to fail. Interpreted through the new framework, this story warns against unity driven by hubris rather than higher purpose. **The breakdown in communication leading to disunity can be seen as an allegory for how cooperative potential is lost when misdirected.** An evolved society must unite under constructive goals (not pride or domination).

## Exodus and Moses – Ethical Cohesion (Exodus 20, etc.)

The formation of Israelite society after the exodus from Egypt is centered on receiving a divine law code (the Ten Commandments and many others). These laws demand moral behavior (honesty, altruism, justice, community welfare) which helped transform a group of former slaves into a cohesive nation. **This reflects an evolutionary step toward a more morally advanced community: survival and prosperity are achieved not just through freedom, but through shared ethical commitments that bind people together.**

## Prophets' teachings on social justice (e.g., Isaiah, Micah)

Biblical prophets repeatedly emphasize justice, compassion, and caring for the vulnerable (orphans, widows, the poor) as core duties of society. For instance, Micah 6:8 famously says what God requires is "to act justly, love mercy, and walk humbly." These injunctions **encourage society to evolve beyond mere tribalism or power struggles**, guiding it toward a cooperative ethos centered on justice and mercy, precisely the traits our evolutionary-spiritual model holds as vital for future progress.

## Parable of the Good Samaritan (Luke 10:25–37)

This parable vividly illustrates principles of altruism, compassion, and cooperation extending beyond familial or ethnic ties. By portraying genuine care provided by an individual considered an outsider, the story **promotes universal empathy, encouraging societies toward greater inclusivity, interconnectedness, and global cooperation.** Such behavior represents a crucial evolutionary step, **guiding humanity** toward expanded social cohesion and unified ethical responsibility.

## Jesus 'Sermon on the Mount (Matthew 5–7)

Jesus 'Sermon on the Mount presents transformative moral principles such as peace, forgiveness, altruism, and moral integrity. **These values form an ethical blueprint that emphasizes communal harmony, emotional resilience, and compassionate interaction.** Embracing and practicing these ideals is essential for creating sustainable, cooperative, and cohesive societies, thereby significantly advancing humanity's evolutionary potential.

## Early Christian (Acts 2:42–47)

The communal lifestyle of early Christians exemplifies collective sharing, cooperation, and mutual support. By voluntarily pooling resources, caring for the vulnerable, and prioritizing collective welfare over individual accumulation, these communities effectively modeled organizational structures that foster group resilience and societal harmony. **Such cooperative living arrangements align closely with evolutionary principles, highlighting optimal strategies for group survival, social stability, and progressive communal development.**

### Quranic Stories

## Adam as Earth's Steward (Quran "The Cow" 2:30–33)

The Quran also begins with the story of Adam, who is taught the names of all things and placed on Earth as a khalifa (steward of God). This role implies **responsible governance of Earth**. Adam's story in the Quran highlights knowledge as well, as God teaches Adam, suggesting **human intellect and learning are key**. The stewardship and pursuit of knowledge entrusted to humanity align with the idea that humans are meant to consciously guide the planet's future, an evolutionary responsibility.

## Story of Prophet Yusuf (Joseph) (Quran "Joseph" 12)

Joseph's saga in the Quran emphasizes foresight, planning, and moral integrity. Notably, Joseph interprets Pharaoh's dream of seven fat and seven lean years and implements a plan that saves an entire region from famine. This narrative illustrates **resilience**, **planning for future** scarcity (an evolutionary advantage), and **using wisdom to benefit society**. It aligns with our concept by showing how **knowledge,**

**cooperation, and ethical leadership**, demonstrated when Joseph forgives and helps those who wronged him, lead to collective survival and prosperity.

### Prophet Shuʿaib and Economic Justice (Quran "Hud" 11:84–95)

Prophet Shuʿaib is sent to a people who were cheating in weights and measures, an act of economic injustice. He calls them to fairness in trade and warns that dishonesty and greed will lead to ruin. This story underscores the **importance of economic cooperation and honesty as foundations of a stable society**. Fair commerce ensures trust and mutual benefit, clearly conducive to a cooperative and sustainable civilization. Societies that evolve greater fairness and reduce exploitation are more likely to endure and thrive.

### Prophet Muhammad's Constitution of Medina

While not a "story" in the Quran, the historical Charter of Medina orchestrated by Prophet Muhammad is a powerful example. It established a multi-tribal, multi-faith community with mutual obligations and **justice among Muslims, Jews, and others** in the city. This early social contract fostered **unity and collective governance across tribal lines.** It reflects the evolutionary leap of **transcending tribalism** to form a pluralistic society bound by common principles, a direct parallel to our model's emphasis on expanding cooperation and social integration.

### Verses Emphasizing Unity (Quran "The Family of Imran" 3:103)

The Quran explicitly urges believers to "hold fast, all together, to the rope of God, and be not divided." This verse, among others, promotes unity and condemns division. **Unity is portrayed as both a spiritual and practical strength**. In our framework, such exhortations can be seen as guiding humanity toward greater collective cohesion, an evolutionary advantage. Societies united in purpose and values can achieve higher levels of organization and face challenges more effectively than fragmented ones.

## Verses on Knowledge and Reflection (Quran "The Family of Imran", "The Groups" 3:191, 39:9)

The Quran repeatedly encourages using reason, observing nature, and seeking knowledge. For example, it asks, "Are those who know, equal to those who do not know?" and praises "those who reflect on the creation of the heavens and the earth." This **emphasis on intellectual development and reflection aligns with the evolutionary trajectory toward advanced intelligence and understanding.** It nudges believers to continually learn and adapt, qualities crucial for long-term evolution and problem-solving.

## Zakat (Charity) and Economic Cooperation (Quran "The Repentance" 9:60)

The Quran mandates Zakat, a form of almsgiving or wealth tax, as one of the Five Pillars of Islam. By institutionalizing charity, the Quran reduces economic inequality and fosters social solidarity. Everyone contributes to a system that **supports the poor and vulnerable, enhancing group survival and harmony**. This practice directly promotes a cooperative societal structure and can be viewed **as an evolutionary mechanism for social cohesion**, ensuring resources circulate to where they are needed for the health of the whole community.

### Potential Divergence or Complexities:

- **Individual Judgment and Salvation**

Possible Divergence:

Some theological interpretations prioritize individual spiritual accountability over collective evolution. However, collective morality and accountability often underpin religious doctrines, which still aligns with cooperative evolution in a broader interpretation.

- **Eschatology (End-Times narratives)**

Complex Alignment:

Apocalyptic narratives sometimes emphasize divine intervention over human cooperation. However, these could be interpreted symbolically as warnings that highlight the urgency and necessity of human cooperation to avoid societal collapse.

In summary, upon evaluating these key spiritual teachings, it is increasingly obvious that **most Biblical and Quranic narratives strongly align with the proposed evolutionary approach**, emphasizing **collective responsibility, social harmony, ethical behavior, stewardship, justice, intellectual and moral evolution, and cooperative survival strategies.** Divergences mainly arise from interpretations emphasizing individualism or apocalyptic passivity; yet broader spiritual readings often resolve these divergences by framing them as instructive warnings rather than inevitable outcomes.

The next step is to look at a series of concepts and directions provided in **historical Christian and Islamic texts** to see how they may fit in with this framework, and where there may be divergence.

# Sacred Evolution

# Part II: Core Virtues

# Chapter 4: Piety

**Piety: Definition and Purpose**

Piety is the virtue of **reverent devotion**, characterized by a profound sense of spiritual responsibility and moral alignment with a higher authority. It involves not only outward religious observance but also inward humility, integrity, and ethical consistency. Piety serves as the **bridge between belief and behavior**, between the sacred and the everyday.

From an evolutionary-spiritual perspective, piety functions as a stabilizing moral force. It cultivates internal regulation, fosters communal solidarity, and encourages individuals to act in accordance with transcendent values, even when external enforcement is absent.

**Piety: Christian Perspective**

In Christianity, piety is often expressed through godliness, a term used in both the Old and New Testaments. It includes **prayer, worship, moral living**, and **compassionate action**, all grounded in a loving relationship with God.

Jesus criticizes hypocritical piety; acts done for show rather than from a sincere heart:

> *"When you pray, do not be like the hypocrites... Truly I tell you, they have received their reward."* (Matthew 6:5)

Instead, He teaches a hidden, sincere, relational piety, summed up in the commandment:

*"Love the Lord your God with all your heart, soul, and mind."* (Matthew
22:37)

In Christian tradition, particularly among the saints, piety is expressed in:

- **Personal prayer and devotion**

- **Charity and humility**

- **Ethical witness**, even at great cost (i.e. living out one's moral
  and spiritual convictions publicly, even when doing so leads to
  suffering, persecution, or death.)

Piety in Christianity is **relational** rather than ritualistic; a response to
God's grace expressed in love, justice, and service.

## Piety: Islamic Perspective

In Islam, the concept of piety is most often conveyed by the term **taqwā**,
commonly translated as **God-consciousness**, **moral vigilance**, or
**spiritual awareness**. It is considered the highest human quality in the eyes
of God:

*"Indeed, the most noble of you in the sight of God is the most God-
conscious of you."* (Qur'an "The Chambers" 49:13)

Taqwā integrates belief, practice, and ethical restraint. A pious person in
Islam is one who:

- **Prays regularly and sincerely**

- **Refrains from sin** not out of fear of punishment, but from awe
  of the Divine

- **Upholds justice** and **honesty** even when no one is watching

The Prophet Muhammad embodied taqwā in all aspects of life. His humility, mercy, consistency, and moral courage are held up as the gold standard of piety. Islam views piety not as separation from the world, but as **active moral engagement rooted in spiritual consciousness**.

**Piety: Does it Align with Future Human Growth and Evolution? (See Appendix 1.0)**

| Criterion | Assessment |
|---|---|
| **1. Communication** | ✓ Piety shapes honest, respectful discourse and **intergroup empathy.** |
| **2. Friction** | ✓ Encourages ethical restraint, forgiveness, and humility. |
| **3. Problem Solving** | ✓ Provides **inner clarity** and consistent moral grounding in complexity. |
| **4. Resilience** | ✓ **Anchors the self** during hardship, uncertainty, and loss. |
| **5. Trust & Cooperation** | ✓ Pious individuals tend to act dependably and altruistically. |
| **6. Adaptability** | ✓ Piety **adapts across cultural and technological change**, as an inner orientation. |
| **7. Pro-Social w/o Reward** | ✓ **Motivated by conscience** and transcendence, not recognition. |

| 8. Functional Health | ✔ Associated with **reduced anxiety, stronger relationships, and lower stress**. |
|---|---|

## Piety: Points of Alignment

- Both Islam and Christianity affirm that true piety is **inward**, sincere, and morally generative.

- Piety integrates **ritual, ethics, and intention**, not merely one's beliefs but one's conduct.

- It is seen as the criteria of achieving success and nobility in the eyes of God, as opposed to worldly criteria such as wealth, lineage or.

## Piety: Points of Divergence or Nuance

- Islam tends to structure piety ritually and prescriptively (e.g., five daily prayers, fasting), while Christianity emphasizes spiritual piety rooted in relationship with God. It emphasizes communion in practice.

- Christianity more strongly critiques public religiosity, whereas Islam celebrates visible markers of faith when sincere.

- Both stress that **inner sincerity is indispensable, but the expressions and institutions of piety differ culturally and historically**.

## Piety Conclusion: Evolutionary Trajectory

Piety is a moral stabilizer and spiritual compass, integrating reverence with responsibility. It orients humans toward **self-discipline, service, and moral vigilance**, reinforcing ethical behavior even when unobserved or unrewarded. Both Islam and Christianity elevate piety as the root of human flourishing, offering a profound alignment between personal growth, communal trust, and spiritual destiny.

In the context of the evolutionary-spiritual framework, piety may serve as a vital catalyst for humanity's next great transition. As we move beyond mere biological survival toward ethical and spiritual refinement, piety, anchored in humility, reverence, and inward accountability, could act as both **a stabilizing force and a directional compass.** It cultivates inner restraint and moral clarity, enabling individuals and communities to **resist destabilizing impulses** and to orient themselves toward **higher forms of cooperation, compassion, and existential purpose**. Far from being a relic of pre-modern religiosity, piety may emerge as an essential trait in shaping the evolutionary trajectory of a species striving not only to persist in the cosmos but also to flourish meaningfully within it.

# Chapter 5: Love

**Love: Definition and Purpose**

Love is the active commitment to the well-being, dignity, and flourishing of another, expressed through compassion, empathy, mercy, and sacrifice. It transcends fleeting emotion or self-interest, forming the core of relational ethics and spiritual authenticity. Love is what moves individuals and societies from simply coexisting to **connecting meaningfully**.

In the evolutionary-spiritual framework, love is not only a personal virtue, it is a necessity for progression of any sort. It binds individuals into families, communities, and moral ecosystems, promoting trust, cohesion, and enduring cooperation.

**Love: Christian Perspective**

Christianity elevates agape, unconditional, self-giving love, as **the greatest of all virtues**:

> *"Now these three remain: faith, hope, and love. But the greatest of these is love."* (1 Corinthians 13:13)

Jesus commands:

> *"Love the Lord your God... and love your neighbor as yourself."* (Matthew 22:37–39)

> *"Love your enemies and pray for those who persecute you."* (Matthew 5:44)

The crucifixion is the defining symbol of divine love: a love that sacrifices, redeems, and transforms. Christian love is:

- **Forgiving**: even in the face of **injustice**

- **Universal**: transcending tribal, social, or national **boundaries**

- **Incarnational**: lived in **service**, healing, and solidarity with the suffering

Love is both a commandment and a lifestyle, rooted in God's nature and reflected in human action.

## Love: Islamic Perspective

In Islam, love (***maḥabbah***) is both divine and moral. God is Al-Wadūd; The Most Loving, whose mercy encompasses all things (Qur'an "The Elevated Place" 7:156). Love is not romanticized, but expressed through mercy, justice, and service.

> *"Indeed, those who believe and do righteous deeds—the Most Merciful will appoint for them love."* (Qur'an "Mary" 19:96)

Islamic love is:

- Relational: love for God, His Messenger, and fellow beings

- Action-based: expressed through **care, protection, forgiveness, and charity**

- Disciplined: guided by **justice, balance, and divine boundaries**

The Prophet Muhammad exemplified love through:

- Tireless **service** to family and strangers

- **Mercy** to enemies

- **Prayer** for those who harmed him

Love in Islam is a sign of spiritual refinement, and a requirement for faith to be complete.

| Criterion | Assessment |
|---|---|
| **1. Communication** | ✔ Deepens **empathy**, nurtures understanding, and **fosters peace**. |
| **2. Friction** | ✔ Heals offense, restrains vengeance, and **bridges division**. |
| **3. Problem Solving** | ✔ Encourages collaborative, compassionate, **long-term solutions**. |
| **4. Resilience** | ✔ Strengthens individuals and communities during crisis. |
| **5. Trust & Cooperation** | ✔ Sustains relationships, institutions, and interdependence. |
| **6. Adaptability** | ✔ Expressed universally, across cultures and circumstances. |
| **7. Pro-Social w/o Reward** | ✔ Exemplifies **altruism**, even when costly. |

| | |
|---|---|
| 8. Functional Health | ✓ Linked to **better mental health, social harmony, and spiritual clarity.** |

## Love: Does it Align with Future Human Growth and Evolution?

### Love: Points of Alignment

- Both traditions **place love at the center** of ethical and spiritual life.

- Love is active, sacrificial, and redemptive. Measured not by words, but by deeds.

- Love is **a sign of maturity, not naivety**; it is discipline in the service of compassion.

### Love: Points of Divergence or Nuance

- Christianity emphasizes love as **grace-driven** and **redemptive**, even for enemies and the unworthy.

- Islam frames love within moral boundaries and reciprocal justice, **prioritizing mercy** but not overlooking accountability.

- Christian beliefs idealize unconditional love, and both Christian beliefs and Islamic love often includes measured consequence with compassion.

### Love Conclusion: Evolutionary Trajectory

Love is the most powerful evolutionary-spiritual force. It turns isolated individuals into **enduring communities, transforms strangers into kin,** and redirects the arc of history from conflict to compassion. Both Islam and Christianity proclaim that **without love, law collapses, rituals hollow, and faith fades.**

In the Evolutionary-Spiritual (ES) framework, love is not merely a moral sentiment or private virtue. Rather, it is a **bio-social mechanism** that has shaped, and must continue to shape, humanity's evolutionary trajectory. From early kin bonding to expansive moral inclusion, love has been the **binding agent of cooperative survival**, from family units to entire civilizations. The ES framework evaluates religious conceptions of love not just for their ethical elegance but for their capacity to cultivate resilient, just, and interdependent societies. The Evolutionary-Spiritual framework suggests that **human evolution is no longer biological alone**, it is now spiritual and ethical as well. **Love becomes a selective pressure**: those communities that master both sacrificial love and structured mercy will be the **most likely to adapt**, endure, and elevate humanity's collective consciousness.

# Chapter 6: Mercy

**Mercy: Definition and Purpose**

Mercy is the **compassionate withholding of punishment** or harm, often in response to wrongdoing or vulnerability. It reflects the **willingness to forgive**, pardon, and uplift, even when justice might allow for retribution. **Mercy is not the opposite of justice**. It is justice tempered by empathy and hope. It allows communities to heal rather than fracture, and individuals to grow **rather than be discarded**.

In the evolutionary-spiritual model, mercy enables systems to **self-correct** without collapse. It allows for reintegration, moral repair, and relational stability, particularly in high-stakes environments of offense or failure.

**Mercy: Christian Perspective**

Mercy is at the **heart of the Gospel**. God is portrayed not just as judge but as a Father of mercies (2 Corinthians 1:3).

Jesus says:

*"Be merciful, just as your Father is merciful."* (Luke 6:36)

*"Blessed are the merciful, for they shall obtain mercy."* (Matthew 5:7)

Mercy in Christian theology is:

- Embodied in Christ: His healing, forgiveness, and ultimate sacrifice

- Expressed through love of enemies, care for the poor, and **non-retaliation**

- A sign of spiritual maturity, manifesting in forgiveness, reconciliation, and grace

The parable of the Prodigal Son (Luke 15) illustrates mercy as restoration, not merely pardon.

**Mercy: Islamic Perspective**

Mercy (*raḥmah*) is a central divine attribute in Islam. Every chapter of the Qur'an (except one) begins with:

*"In the name of God, the Most Gracious, the Most Merciful (Ar-Raḥmān, Ar-Raḥīm)."*

In addition to God's Mercy, the importance of exercising mercy is stressed:

*"If you have power over your enemy, pardon him by way of gratitude for having power over him."* (Nahj al-Balagha, Ali ibn Abi Talib)

God's mercy:

- Encompasses all things (Qur'an "The Elevated Places" 7:156)

- **Outweighs His wrath** (Hadith Qudsi)

- Is the **basis for forgiveness**, provision, and divine patience

Human beings are instructed to embody mercy:

- Toward the weak, the orphan, the animal

- In judgment, commerce, and conflict

- Through forgiveness, **even when wronged**

Mercy is not softness. It is **power restrained by compassion**. It protects the soul from hardening and society from polarizing.

| Criterion | Assessment |
|---|---|
| **1. Communication** | ✔ Opens the way for **healing dialogue** and moral renewal. |
| **2. Friction** | ✔ Prevents **cycles of revenge** and social breakdown. |
| **3. Problem Solving** | ✔ Offers humane responses to complex moral situations. |
| **4. Resilience** | ✔ Allows for personal growth and **reintegration after failure**. |
| **5. Trust & Cooperation** | ✔ Builds social safety nets grounded in compassion. |
| **6. Adaptability** | ✔ Flexible enough to contextualize justice with humanity. |
| **7. Pro-Social w/o Reward** | ✔ Acts of mercy often cost the giver and help the undeserving. |
| **8. Functional Health** | ✔ **Reduces trauma**, strengthens community bonds, and restores dignity. |

**Mercy: Does it Align with Future Human Growth and Evolution?**

**Mercy: Points of Alignment**

- Both Islam and Christianity treat mercy as a **divine trait** to be mirrored in human life.

- Mercy is **not weakness** but moral strength and spiritual refinement.

- In both traditions, mercy restores rather than condemns and elevates the potential of the fallen.

**Mercy: Points of Divergence or Nuance**

- Islam frames mercy within a legal and cosmic order, **balancing mercy with justice**. Each sin, while pardonable, still carries weight and responsibility.

- Christianity tends to emphasize original sin and the need for divine mercy, especially **through Christ's atonement**, and expects believers to extend that mercy universally, even to enemies.

- Islam institutionalizes mercy through rules for restitution, forgiveness, and limiting retribution; Christianity emphasizes interpersonal and divine mercy as the heart of ethical life.

**Mercy Conclusion: Evolutionary Trajectory**

In the evolutionary-spiritual vision, mercy is not the suspension of justice but its elevation and its humanization in service of a higher path. As humanity evolves beyond mere biological survival toward ethical resilience and spiritual maturity, mercy emerges as an indispensable trait. It interrupts cycles of vengeance, prevents social fragmentation, and fosters the kind of trust and compassion that allow communities to endure across generations. **Without mercy, justice can become rigid and retributive; with mercy, it becomes transformative**. Islam and

Christianity both affirm that mercy is not peripheral but foundational: a society that forgives, that sees the Divine even in the broken and the penitent, is not just morally admirable, but it is evolutionarily superior. Such a society does not merely survive. It flourishes, building a future marked by coherence, restoration, and ethical depth.

# Chapter 7: Justice

## Justice: Definition and Purpose

Justice is the consistent commitment to **fairness**, **equity**, and **moral accountability**. It is the virtue that governs how power is exercised, how wrongs are righted, and how dignity is upheld in personal and social relationships. Justice **protects the weak**, **restrains the strong**, and ensures that rights are preserved and responsibilities fulfilled.

In the evolutionary-spiritual context, justice functions as the load-bearing structure of any sustainable society. Without it, trust breaks down, cooperation collapses, and ethical growth is stunted. Justice ensures that communities endure not by force, but by fairness.

## Justice: Christian Perspective

Justice in Christianity is rooted in both the character of God and the life of Jesus, who defended the vulnerable and challenged unjust power:

*"Let justice roll down like waters, and righteousness like an ever-flowing stream."* (Amos 5:24)

Biblical justice includes:

- **Righteousness** (Hebrew: tzedakah): living rightly in relationship with others

- Judicial fairness: **impartiality**, protection for the oppressed

- Restorative ethics: healing broken relationships and systems; **not just for retribution**

Jesus frames justice not as legalism but as mercy and fidelity to God's heart:

*"Woe to you... for you neglect justice and the love of God."* (Luke 11:42)

Christian ethics demand justice in:

- Economic life (condemning exploitation)

- Social life (lifting the poor and stranger)

- Spiritual life (judging oneself before judging others)

**Justice: Islamic Perspective**

Justice (*'adl*) is one of the core values of Islam, inseparable from belief in God. The Qur'an declares:

> *"Indeed, God commands justice, good conduct, and giving to relatives..."* (Qur'an "The Bee" 16:90)

Justice in Islam is:

- Comprehensive: applying to individual actions, family relations, business dealings, and governance

- **Objective**: not swayed by personal interest or tribal loyalty

- Moral and legal: God is Al-'Adl (The Just), and humans are called to mirror divine justice

The Prophet Muhammad said:

> *"The most beloved of people to God on the Day of Judgment will be the just leader."* (Tirmidhi)

Islamic law (**Sharīʿah**), ethical conduct, and social norms all converge on the goal of preserving justice for all, especially orphans, women, minorities, and the poor.

| Criterion | Assessment |
|---|---|
| 1. Communication | ✔ Clarifies expectations and builds transparent systems. |
| 2. Friction | ✔ **Resolves disputes** and reduces resentment and **inequity**. |
| 3. Problem Solving | ✔ Provides frameworks for **accountability** and redress. |
| 4. Resilience | ✔ Sustains order and legitimacy during crises. |
| 5. Trust & Cooperation | ✔ Fair systems increase social trust and long-term cooperation. |
| 6. Adaptability | ✔ Justice-oriented systems can evolve while remaining principled. |
| 7. Pro-Social w/o Reward | ✔ Just behavior often comes at personal cost, yet benefits society. |
| 8. Functional Health | ✔ Correlates with **lower crime, greater equality, and public stability**. |

**Justice: Does it Align with Future Human Growth and Evolution?**

**Justice: Points of Alignment**

- Both Islam and Christianity view justice as **non-negotiable, sacred**, and socially imperative.

- Justice is seen not as abstract law, but as living alignment with divine character.

- The call to justice includes **restraint of power**, elevation of the vulnerable, and **integrity in daily life.**

**Justice: Points of Divergence or Nuance**

- Islam emphasizes principled justice, deeply embedded in community law and governance. Sharīʿah law aims to preserve justice through codified systems.

- Christianity (particularly in the New Testament) emphasizes relational and restorative justice, rooted in love, repentance, and mercy.

- Islam structures justice formally; Christianity often expresses justice through prophetic witness and acts of mercy.

**Justice Conclusion: Evolutionary Trajectory**

Justice is the scaffold of human dignity and cooperative survival. It provides the ethical structure upon which trust, order, and mutual respect can be built. Without it, civilizations regress into cycles of dominance, exploitation, or collapse. With it, even the most diverse and pluralistic societies can flourish, unified not by sameness but by shared standards of fairness and accountability. Both Islam and Christianity elevate justice beyond mere legality. It is the reflection of divine balance within human history. In the evolutionary-spiritual framework, **justice is not simply a moral aspiration but an evolutionary imperative**. The need for fairness and justice is deeply embedded within the human psyche, and for any future adherence, any principle must have it firmly embedded in it. It enables humanity to rise above instinct and impulse, replacing fear and force with conscience and reason. Societies governed by justice, tempered by compassion and aligned with truth, are more resilient, more inclusive, and more capable of advancing along a higher evolutionary trajectory, where **power serves principle and survival serves meaning**.

# Chapter 8: Charity

## Charity: Definition and Purpose

Charity is the voluntary giving of resources, time, or care to those in need, motivated by compassion, justice, and spiritual generosity. It is an active expression of solidarity, grounded in the belief that **human dignity is universal and sacred**, regardless of wealth, status, or merit.

In the evolutionary-spiritual model, charity functions as a **corrective mechanism for inequality**, a binding force for social trust, and a signal of moral maturity. It reflects an ethic of interdependence rather than competition, sustaining the moral economy of any thriving society.

## Christian Perspective

Charity (*caritas*) is considered the greatest of the theological virtues, expressing divine love made visible.

Paul writes:

> "If I give all I possess to the poor... but have not love, I gain nothing. (1 Corinthians 13:3)

Jesus teaches:

> "Give to the one who asks you... whatever you did for the least of these, you did for me." (Matthew 5:42, 25:40)

Christian charity is:

- Rooted in agape: **unconditional**, self-giving love

- Directed especially toward the poor, the sick, the marginalized

- Practiced through almsgiving, hospitality, and acts of mercy

From the early Church's communal economy (Acts 4:32) to Christian humanitarian missions, charity has been both a **moral imperative** and **spiritual discipline.**

**Charity: Islamic Perspective**

Charity is one of the **pillars** of Islam, integral to both worship and ethical life. It is not merely an economic act but a spiritual discipline that fosters compassion, equity, and communal trust.

Zakāt is obligatory charity, calculated as a fixed percentage (typically 2.5%) of certain types of accumulated wealth. It is directed toward:

- The poor and needy

- Those burdened by debt

- Travelers, orphans, and those engaged in public service

Khums is another form of obligatory charity, emphasized particularly in Shia Islam. It requires giving one-fifth (20%) of surplus income annually and is distributed for:

- The descendants of the Prophet (ṣadaat)

- The Imam or, in his absence, qualified religious authorities (marāji')

- Religious, educational, and social welfare causes. Its purpose is to sustain religious leadership and support community welfare in a just and organized manner.

Ṣadaqah is voluntary charity, encompassing acts of kindness both material and non-material: money, service, time, and even simple gestures of goodwill:

> *"Every act of kindness is charity."* (Ṣaḥīḥ Muslim)

The Qur'an states:

> *"You will not attain righteousness until you spend from that which you love." (Qurʾan "The Family of ʿImrān" 3:92)*

Charity in Islam is:

- A form of spiritual **purification**

- A means of wealth redistribution

- A test of sincerity and selflessness

It is not optional generosity. It is sacred **justice in action**.

| Criterion | Assessment |
|---|---|
| **1. Communication** | ✔ Embodies care, empathy, and shared moral language. |
| **2. Friction** | ✔ Eases resentment by **correcting structural inequality**. |

| | |
|---|---|
| **3. Problem Solving** | ✓ Meets **urgent needs** and reinforces shared responsibility. |
| **4. Resilience** | ✓ Builds social **safety nets** and reinforces trust during hardship. |
| **5. Trust & Cooperation** | ✓ Charity deepens reciprocal bonds and long-term solidarity. |
| **6. Adaptability** | ✓ Universally translatable across economies and cultures. |
| **7. Pro-Social w/o Reward** | ✓ Often anonymous and sacrificial, **not transactionally** motivated. |
| **8. Functional Health** | ✓ Reduces poverty, stress, and marginalization. |

**Charity: Does it Align with Future Human Growth and Evolution?**

**Charity: Points of Alignment**

- Both Islam and Christianity view charity as essential to spiritual life and ethical society.

- Charity is not merely a "nice thing to do". It is a spiritual duty, a test of sincerity, and a way to reflect divine mercy in human affairs.

- Each tradition affirms that giving to the poor is giving to God.

**Charity: Points of Divergence or Nuance**

- Islam codifies charity into mandatory (zakat/khums) and voluntary (ṣadaqah) categories, formalizing justice within economic structures.

- Christianity emphasizes charity as Spirit-led and relational, but also established structured giving; typically, of a minimum 10 percent to the church and the poor.

- Islam stresses transparency, equity, and systematic redistribution, while Christianity highlights compassion and voluntary love as internal motive.

**Charity Conclusion: Evolutionary Trajectory**

Charity is the lifeblood of compassionate civilization. It strengthens the weak, humbles the powerful, and repairs the torn fabric of human community. More than a social duty, it is a signal of moral evolution and the capacity to act not for reward, but for the sake of shared dignity and interdependence. Both Islam and Christianity elevate charity to a divine mandate, grounding it not in surplus, but in sacred responsibility. In the evolutionary-spiritual framework, charity is not a luxury of the virtuous; it is a mechanism of collective resilience. Societies that institutionalize generosity create **ethical ecosystems** where cooperation outpaces competition and mutual care becomes adaptive. In such a future, those who give will not merely be honored for their virtue, they will be recognized as essential **catalysts in humanity's ascent** toward moral sustainability, where survival is no longer measured by strength, but by solidarity.

# Chapter 9: Patience

**Definition and Purpose**

Patience is the disciplined capacity to **endure hardship, delay, or provocation** without despair or retaliation. It enables individuals to **withstand pressure** without abandoning virtue, and societies to maintain coherence under stress. Patience is more than passive waiting. It is the active strength to remain morally grounded in adversity.

In an evolutionary-spiritual framework, patience is a **foundational adaptive trait**. It tempers short-term reactions for long-term gains, fosters cooperation in moments of strain, and serves as a moral anchor when clarity or comfort is absent.

**Patience: Christian Perspective**

In Christianity, patience is both a fruit of the Spirit (Galatians 5:22) and a sign of spiritual maturity.

> *"We also glory in our sufferings, because we know that suffering produces perseverance; perseverance, character; and character, hope."*
> (Romans 5:3–4)

Jesus modeled divine patience in:

- Withstanding **temptation**

- Enduring **persecution**

- Absorbing betrayal and crucifixion **without retaliation**

Paul calls believers to endure trials, delay gratification, and wait on God's timing:

> *"Let us run with perseverance the race marked out for us…"* (Hebrews 12:1)

Christian patience involves trust in divine providence, commitment to nonviolence, and the refusal to let suffering obscure love.

### Patience: Islamic Perspective

Patience (*ṣabr*) is one of the most praised qualities in the Qur'an, mentioned more than 90 times.

> *"O you who believe, seek help through patience and prayer. Indeed, God is with the patient."* (Qur'an "The Cow" 2:153)

Patience is required in:

- Worship: remaining committed to daily practice

- Adversity: enduring trials without complaint

- Ethics: restraining anger, vengeance, or dishonesty

The Prophet Muhammad's entire life was a model of *ṣabr*: facing exile, loss, poverty, war, and betrayal without compromising moral integrity.

> *"Whoever practices patience, God will make him patient. No one is given a gift better and more comprehensive than patience."* (Bukhari)

Patience is not passivity. It is active **spiritual resistance to despair** and moral constancy under fire.

| Criterion | Assessment |
|---|---|
| **1. Communication** | ✔ Enables thoughtful, **non-reactive** dialogue and mutual understanding. |
| **2. Friction** | ✔ **Reduces impulsive conflict, vengeance, and escalation.** |
| **3. Problem Solving** | ✔ Encourages endurance and clarity in complex, long-term challenges. |
| **4. Resilience** | ✔ Strengthens inner will and group stability under prolonged pressure. |
| **5. Trust & Cooperation** | ✔ Demonstrates commitment and reliability in communal life. |
| **6. Adaptability** | ✔ Supports **strategic delay and emotional flexibility** in uncertain environments. |
| **7. Pro-Social w/o Reward** | ✔ Often practiced in secret or under great cost. |
| **8. Functional Health** | ✔ Correlated with lower stress, greater focus, and better emotional health. |

**Patience: Does it Align with Future Human Growth and Evolution?**

**Patience: Points of Alignment**

- Both Islam and Christianity exalt patience as a **form of moral fortitude** and spiritual trust.

- Patience is seen as necessary for enduring suffering without becoming bitter and navigating injustice without becoming unjust.

- Each tradition ties patience to divine companionship and eternal reward.

## Patience: Points of Divergence or Nuance

- Islam often frames patience as **active spiritual training**, part of the broader struggle (*jihad al-nafs*), and a form of **moral resistance**.

- Christianity places patience within the arc of **redemptive suffering**, emphasizing grace and spiritual formation through trial.

- Islamic patience is closely connected to discipline and communal order, while Christian patience is often personal, inward, and linked to hope in Christ's return and the final judgement.

## Patience Conclusion: Evolutionary Trajectory

Patience is a **slow-burning strength**. It is an evolutionary virtue without which all other moral capacities falter under strain. It is the ligament of ethical endurance, holding integrity in place when pain, uncertainty, or temptation threaten collapse. Islam and Christianity alike teach that the patient do not merely survive adversity; they are transformed by it. In the evolutionary-spiritual arc, patience is the mechanism by which civilizations retain their ethical compass through disruption, and individuals **deepen rather than fracture** in the face of trial. It allows for growth that is sustainable, wisdom that is hard-earned, and **virtue that is stress-tested**. Societies that cultivate patience acquire not only inner resilience but long-term cohesion. In such a future, patience is no longer viewed as passive restraint, but as a dynamic force of moral continuity, quietly shaping a humanity capable of enduring hardship without losing its soul.

# Chapter 10: Truthfulness

**Truthfulness: Definition and Purpose**

Truthfulness is the unwavering **commitment to speak, act, and live in alignment with reality**, without distortion, deception, or manipulation. It encompasses not only honesty in words but integrity in conduct, consistency of character, and faithfulness to one's conscience. Truthfulness is the foundation of trust, justice, and moral identity.

In an evolutionary-spiritual framework, truthfulness acts as a social stabilizer and ethical compass. It enables transparent cooperation, reduces suspicion and breakdown, and cultivates a shared moral ecosystem. It is essential not only for individual virtue, but for civilizational coherence.

**Truthfulness: Christian Perspective**

Truth is central to Christian theology. Jesus declares:

> *"I am the way, the truth, and the life."* (John 14:6)

Paul writes:

> *"Therefore each of you must put off falsehood and speak truthfully to your neighbor..."* (Ephesians 4:25)

Truthfulness includes:

- **Speaking honestly**

- **Living consistently with one's values**

- **Bearing witness to divine reality**

Jesus condemned hypocrisy and praised integrity:

*"Let your 'Yes' be 'Yes,' and your 'No,' 'No'; anything beyond this comes from the evil one."* (Matthew 5:37)

Truth is not just ethical. It is **spiritual clarity**, aligning the soul with the divine.

**Truthfulness: Islamic Perspective**

Truthfulness (*ṣidq*) is a defining attribute of the believer and of the Prophet Muhammad himself, who was called Al-Ṣādiq (The Truthful) even before revelation.

*"O you who believe, fear God and be with those who are truthful."* (Qur'an "The Repentance" 9:119)

Truthfulness in Islam is:

- A **moral obligation**: even under personal risk or social cost

- A **form of worship** and spiritual elevation

- A sign of **taqwā** (God-consciousness)

The Prophet said:

*"Truthfulness leads to righteousness, and righteousness leads to Paradise."* (Bukhari & Muslim)

Lying, deceit, and hypocrisy (*nifāq*) are viewed as **severe moral failures**. In fact, false testimony is among the gravest sins, often equated with idolatry and murder.

| Criterion | Assessment |
|---|---|
| 1. Communication | ✔ Enables trust, clarity, and **deep interpersonal understanding.** |
| 2. Friction | ✔ Minimizes **suspicion, manipulation, and reactive conflict.** |
| 3. Problem Solving | ✔ Anchors cooperation in **shared reality** and honest dialogue. |
| 4. Resilience | ✔ Builds integrity-based endurance and **self-coherence** under pressure. |
| 5. Trust & Cooperation | ✔ Truthful individuals are more trustworthy and dependable. |
| 6. Adaptability | ✔ Promotes ethical recalibration and continuous moral learning. |
| 7. Pro-Social w/o Reward | ✔ Truthfulness is often maintained without recognition or benefit. |
| 8. Functional Health | ✔ **Reduces cognitive dissonance** and relational breakdown. |

**Truthfulness: Does it Align with Future Human Growth and Evolution?**

**Truthfulness: Points of Alignment**

- Both Islam and Christianity define truthfulness as essential to spiritual health and social trust.

- Truth is not merely factual. It is **faithfulness to what is just**, right, and holy.

- Both traditions teach that the truthful are beloved by God and vital to society's moral spine.

## Truthfulness: Points of Divergence or Nuance

- Islam emphasizes truthfulness **within legal, commercial, and communal contexts**, with an emphasis on testimony and transactional integrity.

- Christianity often frames truthfulness through the scriptures, as a **reflection of God's unchanging nature.**

- Islam permits minor concealment to protect life or mend relations (e.g., in peacemaking), while Christianity leans toward total moral clarity, even at great cost.

## Truthfulness Conclusion: Evolutionary Trajectory

Truthfulness is the cognitive and moral currency of civilization, an evolutionary necessity as much as a spiritual imperative. Where truth is honored, trust flourishes, cooperation strengthens, and systems endure. **Where it is forsaken, societies fracture under the weight of suspicion, manipulation, and decay.** Both Islam and Christianity affirm that truth is not an optional ideal but the ground upon which justice, community, and divine connection are built. In the evolutionary-spiritual framework, **truthfulness is the trait that allows human beings to transcend tribal instinct and cognitive distortion**, enabling shared knowledge, moral transparency, and coherent decision-making. As humanity advances into an era clouded by misinformation and digital illusion, **truthfulness will not merely be virtuous; it will be adaptive**. It will distinguish those fit to nurture sustainable civilizations from those destined to undermine them.

In this future, **the honest will be the architects of trust-based evolution,** guiding humanity through complexity with clarity and integrity.

# Chapter 11: Hospitality

**Hospitality: Definition and Purpose**

Hospitality is the open-hearted **act of welcoming, caring for, and honoring guests, strangers**, or those in need. It is a virtue that extends beyond etiquette to become an expression of empathy, humility, and moral inclusion. At its best, hospitality **creates sacred space for the unfamiliar**, a relational bridge where **fear dissolves and trust is born**.

In the evolutionary-spiritual context, hospitality reflects mature social intelligence. It neutralizes threats, fosters intercultural exchange, and signals moral strength. It transforms "the other" into "the guest," thereby converting risk into relationship.

**Hospitality: Christian Perspective**

Hospitality is central to Christian ethics, modeled on Jesus 'radical openness to all, especially the marginalized.

> *"I was a stranger and you welcomed me."* (Matthew 25:35)

The New Testament teaches:

> *"Do not forget to show hospitality to strangers, for by so doing some have entertained angels unaware."* (Hebrews 13:2)

Christian hospitality includes:

- Receiving the stranger as one would receive Christ

- Creating inclusive communities where **none are rejected**

- Expressing compassion through **food, shelter, presence, and kindness**

Jesus dined with outcasts, healed foreigners, and emphasized that true hospitality is extended to those who cannot repay. The early Church practiced communal meals and care for travelers as hallmarks of faith.

**Hospitality: Islamic Perspective**

Hospitality (***karam* or *dhiyāfah***) is a highly esteemed trait in Islam, rooted in the prophetic tradition and Arab cultural ethics.

The Prophet Muhammad said:

> *"Whoever believes in God and the Last Day should honor his guest."*
> (Bukhari & Muslim)

Key principles of Islamic hospitality include:

- Welcoming guests without hesitation, **offering the best** of what one has

- Providing three days of generosity without expectation of return

- Viewing hospitality as a sign of taqwā (God-consciousness) and an avenue to divine reward

The Qur'an highlights Abraham's immediate care for his unknown visitors:

> *"And he hastened to bring a fat roasted calf..."* (Qur'an "The Scattering Winds" 51:26)

Hospitality in Islam is not merely social courtesy. It is a **sacred duty** and a reflection of **moral nobility**.

| Criterion | Assessment |
|---|---|
| 1. Communication | ✔ Creates openness across lines of difference. |
| 2. Friction | ✔ Reduces **suspicion** and **hostility** toward the "other." |
| 3. Problem Solving | ✔ Builds trust and cooperation through reciprocal exchange. |
| 4. Resilience | ✔ **Strengthens communal bonds** during crisis or displacement. |
| 5. Trust & Cooperation | ✔ Transforms stranger interactions into loyal relationships. |
| 6. Adaptability | ✔ **Effective in all cultures**, especially where social cohesion is fragile. |
| 7. Pro-Social w/o Reward | ✔ Often extended with no expectation of return. |
| 8. Functional Health | ✔ Linked to **reduced isolation**, increased **empathy**, and stronger networks. |

**Hospitality: Does it Align with Future Human Growth and Evolution?**

**Hospitality: Points of Alignment**

- Both Islam and Christianity view hospitality as a **moral test and spiritual act**, not merely a social custom.

- Welcoming the guest is seen as welcoming God's presence, and as an **opportunity to manifest virtue.**

- Hospitality is framed as an **obligation of faith**, not a luxury of convenience.

## Hospitality: Points of Divergence or Nuance

- Islam formalizes hospitality with specific etiquettes, time frames, and rewards, particularly in Bedouin and prophetic traditions.

- Christianity often characterizes hospitality as a symbol of divine grace and radical inclusion, especially in Jesus 'ministry.

- Islam emphasizes honoring the guest, while Christianity often highlights identifying with the guest, especially when they are marginalized.

## Hospitality Conclusion: Evolutionary Trajectory

Hospitality is an ancient instinct and **a future-oriented virtue**, a social technology that extends trust beyond the bounds of blood and tribe. It opens the tent to the stranger, declaring that welcome is stronger than fear, and that dignity belongs to all, not just the familiar. In both Islam and Christianity, the host is a vessel of divine generosity, and the guest is a bearer of sacred responsibility. Within the evolutionary-spiritual framework, hospitality becomes more than etiquette; it is a **transformative force that expands the moral circle, disarms hostility, and fosters adaptive networks of solidarity**. By inviting the outsider in, societies build the cognitive and ethical architecture needed for inclusive cooperation, cultural resilience, and long-term peace. In this vision of the future, hospitality is not a relic of the past; **it is a survival trait for civilizations seeking to evolve not just technologically, but ethically and spiritually**. It is how strangers become kin, and **how humanity becomes whole.**

# Chapter 12: Modesty

**Modesty: Definition and Purpose**

Modesty is the virtue of **measured self-presentation**, grounded in humility, dignity, and ethical restraint. It involves a conscious effort to avoid excess in appearance, behavior, or speech, fostering respect for the self and others. Modesty protects what is sacred, cultivates moral clarity, and creates boundaries that preserve the human and the divine from trivialization.

In an evolutionary-spiritual framework, **modesty serves as a social signal of self-regulation, trustworthiness, and maturity.** It reduces destructive competition, enhances mutual respect, and helps maintain a balanced, dignified public sphere.

**Modesty: Christian Perspective**

Christian modesty is a fruit of humility, reverence, and grace. It is tied not only to dress but to character and inner disposition:

*"Clothe yourselves with humility toward one another."* (1 Peter 5:5)

Paul teaches:

*"Your beauty should not come from outward adornment... but from your inner self, the unfading beauty of a gentle and quiet spirit."* (1 Peter 3:34)

Modesty in Christian thought includes:

• Gentleness of speech

- Simplicity of lifestyle

- Avoidance of vanity or **public self-exaltation**

- Sexual modesty as an expression of self-respect and spiritual discipline

Jesus modeled modesty in power, refusing ostentation even in miracles, and taught that greatness comes through service and humility.

**Modesty: Islamic Perspective**

Modesty (*ḥayā*) is considered a core virtue in Islamic ethics, directly linked to faith itself:

> *"Every religion has a distinct characteristic, and the characteristic of Islam is modesty."* (Ibn Mājah)

The Prophet Muhammad said:

> *"Modesty brings nothing but good."* (Bukhari)

Modesty in Islam includes:

- **Physical modesty**: appropriate dress and behavior for both men and women

- **Behavioral modesty**: humility in speech, avoidance of arrogance, and deference to others 'dignity

- **Spiritual modesty**: recognizing one's dependence on God and refraining from self-righteousness

The Qur'an advises:

*"Tell the believing men to lower their gaze and guard their private parts... And tell the believing women..."* (Qur'an "The Light" 24:30–31)

Modesty is not repression. It is a form of **moral beauty and internal nobility**.

| Criterion | Assessment |
|---|---|
| **1. Communication** | ✔ Encourages respectful interactions and **reduces social signaling** pressure. |
| **2. Friction** | ✔ Lowers **vanity-driven competition** and **objectification**. |
| **3. Problem Solving** | ✔ Creates a focus on **substance over spectacle** in discourse. |
| **4. Resilience** | ✔ Promotes **emotional and social balance** through internal focus. |
| **5. Trust & Cooperation** | ✔ Signals **ethical restraint** and moral reliability. |
| **6. Adaptability** | ! Requires careful contextual interpretation across cultures and eras. |
| **7. Pro-Social w/o Reward** | ✔ Often practiced privately and motivated by conscience. |

| | |
|---|---|
| **8. Functional Health** | ✔ Correlates with lower stress, reduced self-objectification, and **respect-based relationships**. |

## Modesty: Does it Align with Future Human Growth and Evolution?

### Modesty: Points of Alignment

- Both Islam and Christianity treat modesty as an **expression of reverence for oneself**, for others, and for God.

- Modesty reflects **inner discipline**, not just external restraint.

- The modest person is not invisible but purposefully restrained, grounded in deeper values than self-display.

### Modesty: Points of Divergence or Nuance

- Islam codifies modesty in explicit legal and behavioral guidelines, especially regarding gender interaction and dress.

- Christianity, especially in modern interpretations, tends to internalize modesty as an attitude of the heart, with fewer legalistic boundaries.

- Islamic modesty is communal and visible; Christian modesty often emphasizes personal humility and spiritual quietness.

### Modesty Conclusion: Evolutionary Trajectory

Modesty is a quiet but evolutionary strength, resisting the gravitational **pull of ego, exhibition, and excess** in favor of dignity, depth, and inner clarity. It safeguards the self not by hiding it, but by grounding it in purpose rather than performance. Both Islam and Christianity uphold modesty not to suppress the human being, but **to refine and elevate** the soul above spectacle and self-promotion. In the evolutionary-spiritual framework, modesty emerges as a trait that fosters sustainable identity, protects against the disintegration of selfhood in hyper-competitive cultures, and nurtures communities where humility strengthens cooperation. As humanity evolves toward higher forms of consciousness and collective responsibility, modesty will not disappear; it will be **rediscovered as a countercultural force that anchors identity in substance, shields what is sacred**, and cultivates civilizations where restraint becomes a form of wisdom, and silence becomes **a form of strength.**

# Chapter 13: Gratitude

### Gratitude: Definition and Purpose

Gratitude is the **mindful recognition of goodness** received, accompanied by a posture of thankfulness, humility, and moral responsiveness. It transforms entitlement into appreciation, scarcity into abundance, and isolation into connection. Gratitude is not merely a feeling; it is a way of seeing and being, forming the emotional and ethical foundation of contentment, generosity, and hope.

In an evolutionary-spiritual context, gratitude functions as a **psychological and social regulator**. It fosters cooperation, stabilizes mood, and strengthens relational bonds, making it an essential trait for both individual resilience and communal thriving.

### Gratitude: Christian Perspective

Gratitude is a central theme of Christian life and theology, closely tied to **grace and worship**. The Greek word *eucharistia*, meaning "thanksgiving," gives its name to the Eucharist, the highest Christian act of worship.

Paul writes:

> *"Give thanks in all circumstances, for this is God's will for you in Christ Jesus."* (1 Thessalonians 5:18)

Christian gratitude includes:

- Thankfulness to God for salvation, daily sustenance, and providence

- Response to grace, not mere reaction to comfort

- **An attitude of joyful dependence and humility**, even in suffering

Jesus models gratitude in feeding the multitudes, healing the sick, and praying before meals, even before His death. Gratitude in Christianity is a way of abiding in divine love, even amidst hardship.

**Gratitude: Islamic Perspective**

Gratitude (*shukr*) is a core ethical and spiritual virtue in the Qur'an. Believers are repeatedly commanded to recognize and give thanks for:

- Life, provision, health, and guidance

- Both ease and adversity, as tests and opportunities for reflection

*"If you are grateful, I will surely increase you [in favor]"*. (Qur'an "Abraham" 14:7)

Gratitude in Islam is expressed through:

- **Verbal praise**: *Alḥamdulillāh* ("All praise is due to God")

- **Action**: using blessings responsibly, sharing with others

- **Spiritual attentiveness:** remembering God in every state

The Prophet Muhammad was described as "a grateful servant", even praying long hours after a day of service and hardship. Gratitude, in Islam, is not passive. It is a form of worship and worldview.

| Criterion | Assessment |
|---|---|
| **1. Communication** | ✔ Fosters affirming dialogue and relational reciprocity. |
| **2. Friction** | ✔ Diminishes **resentment, envy, and entitlement.** |
| **3. Problem Solving** | ✔ Reframes **challenges as opportunities**; promotes **optimism**. |
| **4. Resilience** | ✔ Enhances emotional strength and recovery from adversity. |
| **5. Trust & Cooperation** | ✔ Builds relational warmth, loyalty, and mutual respect. |
| **6. Adaptability** | ✔ Enables flexible, appreciative responses across life contexts. |
| **7. Pro-Social w/o Reward** | ✔ Cultivates generosity, often without demand for reciprocation. |
| **8. Functional Health** | ✔ Linked to lower depression, greater life satisfaction, and better sleep. |

**Gratitude: Does it Align with Future Human Growth and Evolution?**

**Gratitude: Points of Alignment**

- Both Islam and Christianity see gratitude as **a pillar of faith and a path to joy**.

- Gratitude is **more than a feeling**. It is a disposition of the soul that reshapes how one relates to God, self, and others.

- Both traditions link gratitude to action, responsibility, and ethical stewardship.

**Gratitude: Points of Divergence or Nuance**

- Islam places strong emphasis on gratitude as **spiritual accountability**—a response to blessings as a test.

- Christianity often connects gratitude to grace and unearned favor, highlighting divine generosity beyond merit.

- In Islamic theology, **ingratitude (*kufr*) is closely linked to spiritual blindness**; in Christianity, ingratitude is often a sign of pride or spiritual numbness.

**Gratitude Conclusion: Evolutionary Trajectory**

Gratitude is not mere sentiment; it is an evolutionary virtue, a spiritually rooted survival strategy that reinforces trust, reciprocity, and emotional resilience. It expands our capacity for joy while tempering reactivity and entitlement, anchoring us in a mindset of abundance rather than scarcity. Both Islam and Christianity teach that the grateful soul is the awakened soul, attuned to grace, receptive to the unseen, and inclined to give as it has received. In the evolutionary-spiritual framework, gratitude is not just a moral nicety; it is a foundational trait for civilizations aspiring to endure. It **fosters psychological equilibrium, deepens social bonds**, and cultivates the humility necessary for living in balance with the Earth and with one another. **Gratitude transforms consumption into stewardship and competition into kinship.** In the unfolding arc of human evolution, **gratitude becomes a form of reverent intelligence**, a recognition that survival is not simply about strength, but about remembering what sustains us and honoring it with care.

# Chapter 14: Contentment

**Contentment: Definition and Purpose**

Contentment is the inner state of **peaceful sufficiency**, free from restlessness, envy, or obsessive desire for more. It reflects a posture of acceptance, gratitude, and emotional balance, allowing individuals to live with dignity, serenity, and ethical focus in a world driven by scarcity and comparison.

In the evolutionary-spiritual context, contentment functions as a **brake against destabilizing ambition**. It allows for **long-term thinking**, interpersonal trust, and stable self-worth—traits essential for sustaining moral coherence and collective flourishing in diverse, dynamic societies.

**Contentment: Christian Perspective**

Contentment is emphasized throughout the New Testament, especially in the teachings of Paul:

> *"I have learned to be content whatever the circumstances… whether in plenty or in want."* (Philippians 4:11–12)

Christian contentment is grounded in:

- Trust in God's providence, not worldly control

- Simplicity and **detachment from material excess**

- The belief that spiritual wealth outweighs material gain:

*"Godliness with contentment is great gain."* (1 Timothy 6:6)

Jesus teaches contentment through:

- Encouraging daily trust: *"Give us this day our daily bread"*

- Warning against storing up treasures on earth (Matthew 6:19–21)

- Affirming that true riches lie in relationship with God, not in accumulation

**Contentment: Islamic Perspective**

Contentment (*qanā'ah*) is a revered quality in Islamic tradition, often described as the key to inner wealth. The Prophet Muhammad said:

> *"Richness is not having many possessions, but richness is being content with oneself."* (Bukhari & Muslim)

The Qur'an echoes this ethic:

> *"So do not covet what We have bestowed upon some of them…"* (Qur'an 20:131)

Contentment in Islam is:

- A sign of trust in God's decree (taqdīr) and reliance (tawakkul)

- A shield against greed, consumerism, and moral compromise

- An expression of gratitude, self-respect, and ethical detachment

While Islam encourages striving (*kasb*) for lawful livelihood and excellence, it insists that ultimate satisfaction comes from the heart, not material abundance.

| Criterion | Assessment |
|-----------|------------|
| **1. Communication** | ✔ **Reduces status signaling**, enhances sincerity and empathy. |
| **2. Friction** | ✔ Lessens competition, envy, and societal polarization. |
| **3. Problem Solving** | ! May temper material innovation but enhances moral clarity and resource sustainability. |
| **4. Resilience** | ✔ Fortifies inner peace against external fluctuation. |
| **5. Trust & Cooperation** | ✔ Content individuals are more generous, less manipulative. |
| **6. Adaptability** | ✔ Fosters flexibility and **calm in unpredictable environments**. |
| **7. Pro-Social w/o Reward** | ✔ Frees the self to act morally without desire for personal gain. |
| **8. Functional Health** | ✔ Linked to reduced anxiety, lower cortisol, and higher well-being. |

**Contentment: Does it Align with Future Human Growth and Evolution**

**Contentment: Points of Alignment**

- Both Islam and Christianity celebrate contentment as a mark of spiritual maturity and moral health.

- Contentment resists the idolatry of acquisition, instead rooting value in the soul's alignment with the Divine.

- It serves as a form of peaceful protest against greed and status-based identity.

**Contentment: Points of Divergence or Nuance**

- Islam integrates contentment with **cosmic trust**—seeing provision as divinely appointed, with efforts framed by gratitude and spiritual realism.

- In Christianity, sufficiency is found in union with Christ, not circumstances.

- Both traditions value striving—but only when moderated by inner stillness and spiritual detachment.

**Contentment Conclusion: Evolutionary Trajectory**

Contentment is not complacency; it is evolved clarity, the cultivated wisdom to recognize sufficiency in a world driven by endless appetite. It guards the soul against the disintegration caused by constant craving and orients the self toward deeper joy, relational balance, and morally grounded priorities. Both Islam and Christianity affirm that the contented person is not stunted but spiritually mature, fully alive, anchored in what endures rather than what dazzles. In the evolutionary-spiritual framework, **contentment emerges as a stabilizing force** in the arc of human development. It moderates consumption, fosters psychological resilience, and lays the groundwork for sustainable communities built on gratitude rather than greed. As humanity faces the twin threats of ecological exhaustion and spiritual dislocation, contentment offers an antidote, **a way of living that honors limits not as constraints but as portals to greater** meaning. In the future we must grow into, **contentment will not be a retreat from ambition; it will be its ethical reformation**.

# Chapter 15: Diligence

## Diligence: Definition and Purpose

Diligence is the virtue of **persistent, focused, and ethically motivated effort**. It reflects a commitment to completing tasks with care, excellence, and integrity, even when effort is difficult, or recognition is absent. Diligence is not simply busyness—it is purposeful endurance toward meaningful and responsible goals.

In the evolutionary-spiritual model, **diligence supports long-term survival, ethical growth, and societal trust**. It balances vision with perseverance and ensures that good intentions are followed through with disciplined action.

## Diligence: Christian Perspective

In Christianity, diligence reflects faithful stewardship of time, talent, and calling. Paul urges believers:

> *"Whatever you do, work at it with all your heart, as working for the Lord."* (Colossians 3:23)

Diligence is expressed through:

- **Consistency** in prayer and service

- **Commitment** to spiritual growth and moral labor

- **Perseverance** through adversity:

*"Let us not grow weary in doing good..."* (Galatians 6:9)

The parables of the talents (Matthew 25) and wise virgins (Matthew 25:1–13) illustrate that diligence is rewarded—not just in productivity, but in readiness, character, and trustworthiness.

**Diligence: Islamic Perspective**

Diligence is a sign of sincerity and spiritual maturity in Islam. God commands believers to work with intention and integrity:

> *"Say: Work, for God will see your deeds, and [so will] His Messenger and the believers..."* (Qur'an "The Repentance" 9:105)

The Prophet Muhammad emphasized:

> *"God loves that when one of you does a deed, he does it with excellence (iḥsān)."* (Bayhaqi)

Diligence in Islam appears in:

- **Daily acts of worship** (e.g., prayer, fasting, charity)

- Striving for **knowledge** (*ṭalab al-ʿilm*)

- Earning **lawful livelihood** (ḥalāl rizq)

- **Service** to others with reliability and care

Laziness (*kasl*) is warned against as spiritually corrosive, while steady effort is viewed as a form of ongoing worship.

| Criterion | Assessment |
|---|---|
|  |  |

| | |
|---|---|
| **1. Communication** | ✔ Builds reliability and integrity in social exchange. |
| **2. Friction** | ✔ Reduces resentment, dependency, and communal imbalance. |
| **3. Problem Solving** | ✔ Enables sustained inquiry, innovation, and complex solutions. |
| **4. Resilience** | ✔ Strengthens character and perseverance under stress. |
| **5. Trust & Cooperation** | ✔ Diligent individuals are trusted contributors. |
| **6. Adaptability** | ✔ Supports **consistent effort even under changing conditions**. |
| **7. Pro-Social w/o Reward** | ✔ Motivated by inner discipline, not recognition. |
| **8. Functional Health** | ✔ Associated with better focus, time management, and self-esteem. |

**Diligence: Does it Align with Future Human Growth and Evolution?**

**Diligence: Points of Alignment**

- Both Islam and Christianity affirm diligence as a **moral virtue and spiritual path**.

- Diligence is a sign of faith in action—a refusal to let ease or fatigue derail one's moral purpose.

- Both link diligence to service, worship, and leadership, not merely productivity.

**Diligence: Points of Divergence or Nuance**

- Islam embeds diligence into **daily worship cycles** and **ethical labor laws**, emphasizing regular, balanced exertion.

- Christianity highlights grace-infused labor, where diligence is not only human effort, but cooperation with divine calling.

- Both traditions affirm that effort must be tempered with humility, recognizing that results are in God's hands, but effort is in ours.

**Diligence Conclusion: Evolutionary Trajectory**

Diligence is the ethical engine of evolutionary progress, the capacity to carry moral intention through to lasting impact. It transforms vision into embodiment, potential into purpose, and conviction into culture. Civilizations stagnate not from lack of ideals, but from the erosion of follow-through; individuals unravel not from ignorance, but from the slow corrosion of will. Islam and Christianity both elevate diligence as a sacred form of devotion, an act of worship, stewardship, and trust in divine timing. Within the evolutionary-spiritual framework, diligence becomes more than personal discipline; **it is a collective force that binds generations in the work of shaping a just and enduring future**. It protects truth from abstraction, beauty from decay, and justice from inertia. In the moral trajectory of human evolution, diligence is the virtue that ensures that the sacred is not only envisioned but also realized, refined, and preserved. **It is how ethical civilizations are not merely imagined but built brick by mindful brick.**

# Chapter 16: Sincerity

## Sincerity: Definition and Purpose

Sincerity is the **alignment of intention, speech, and action**, free from hypocrisy, manipulation, or self-deception. It is the virtue of moral transparency, where what one does and what one means are fully integrated. Sincerity builds trust, integrity, and inner clarity, forming the foundation for authentic faith and ethical relationships.

In the evolutionary-spiritual model, **sincerity serves as a reliability signal**—demonstrating moral consistency and enabling deep cooperation. It sustains personal integrity and public credibility, especially in complex or pluralistic settings where motives matter more than appearances.

## Sincerity: Christian Perspective

Sincerity is central to the teachings of Jesus, who condemned outward religiosity that lacked internal authenticity:

> *"These people honor me with their lips, but their hearts are far from me."* (Matthew 15:8)

Paul writes:

> *"Let love be sincere. Hate what is evil; cling to what is good."* (Romans 12:9)

Christian sincerity includes:

- Moral integrity: **walking in truth even when unobserved**

- Relational authenticity: **loving without pretense** or manipulation

- Spiritual transparency: honest prayer, confession, and humility

In Christian ethics, sincerity is tied to grace, where the believer's heart is made new, and actions flow from inner transformation rather than outward performance.

## Sincerity: Islamic Perspective

Sincerity (*ikhlāṣ*) is the essence of worship and the prerequisite for divine acceptance.

> *"And they were not commanded except to worship God, being sincere to Him in religion…"* (Qur'an "The Clear Proof" 98:5)

The Prophet Muhammad taught:

> *"Actions are judged by intentions…"* (Bukhari & Muslim)

Sincerity manifests in:

- **Pure intention** (niyyah) before every act

- **Avoiding ostentation** (riy'ā), especially in religious practice

- **Consistency** between private devotion and public behavior

Islamic spirituality, particularly in Sufi traditions, emphasizes purification of the heart (tazkiyah) as a path to sincerity—where even good actions are meaningless without truthful motive.

| Criterion | Assessment |
| --- | --- |

| 1. Communication | ✔ Enables **trustful dialogue** and ethical clarity. |
|---|---|
| 2. Friction | ✔ Reduces **duplicity**, **distrust**, and performative conflict. |
| 3. Problem Solving | ✔ Encourages honest feedback and ethical collaboration. |
| 4. Resilience | ✔ Builds internal stability and spiritual alignment. |
| 5. Trust & Cooperation | ✔ Deepens interpersonal and communal **trust**. |
| 6. Adaptability | ✔ Honest intention fosters thoughtful moral recalibration. |
| 7. Pro-Social w/o Reward | ✔ Rooted in conscience, not applause. |
| 8. Functional Health | ✔ Reduces anxiety, moral fatigue, and **cognitive dissonance**. |

**Sincerity: Does it Align with Future Human Growth and Evolution?**

**Sincerity: Points of Alignment**

- Both Islam and Christianity regard sincerity as indispensable to genuine virtue.

- Without sincerity, even good deeds lose their spiritual and moral value.

- Sincerity transforms religion from ritual performance into authentic encounter.

## Sincerity: Points of Divergence or Nuance

- Islam integrates sincerity into daily legal, ritual, and moral intention, especially through *niyyah* and the rejection of *riy'ā* (showing off).

- Christianity frames sincerity as a fruit of grace and inner transformation, often tied to spiritual rebirth.

- Both traditions condemn hypocrisy, but Islam codifies its rejection into formal religious ethics, while Christianity often narrates it through parables and prophetic critique.

## Sincerity Conclusion: Evolutionary Trajectory

In the evolutionary-spiritual framework, sincerity is more than personal authenticity; it is a crucial evolutionary trait essential for the sustainable growth of human societies. Sincerity aligns intention with action, creating transparent relationships built on mutual trust and reliability, qualities indispensable for adaptive cooperation. As humanity navigates increasingly complex social, ecological, and technological challenges, sincerity will act as the ethical compass **guiding our collective choices toward genuine collaboration rather than competitive fragmentation**. Both Islam and Christianity emphasize sincerity as foundational, recognizing that **no civilization can endure when actions diverge from professed values**. On the evolutionary path ahead, sincerity will foster communities resilient enough to face crises openly, relationships strong enough to sustain collective purpose, and institutions trustworthy enough to command enduring loyalty. Ultimately, sincerity ensures that the moral vision humanity professes becomes the lived reality it embodies, driving forward not just survival, but **meaningful, conscious evolution**.

# Chapter 17: Moderation

**Moderation: Definition and Purpose**

Moderation is the disciplined practice of **balance, restraint, and thoughtful proportion** in all aspects of life, including beliefs, behaviors, consumption, and emotional responses. It avoids both excess and deficiency, cultivating a harmonious integration of virtue, reason, and sustainability.

In the evolutionary-spiritual framework, moderation ensures individual well-being, social stability, and long-term adaptability. It **tempers extremes, prevents burnout or radicalization**, and preserves room for reflection, dialogue, and evolution, especially in morally or culturally diverse environments.

**Moderation: Christian Perspective**

Christianity also elevates moderation as a fruit of self-control, humility, and wisdom.

Paul urges:

> *"Let your moderation be known to all."* (Philippians 4:5, KJV)

Moderation is practiced through:

- Temperance: **self-control** in food, drink, speech, and sexual ethics

- **Simplicity**: avoidance of materialism and moral pride

- Balance of grace and truth: **avoiding judgmentalism** while not abandoning moral standards

Jesus warns against **legalistic extremism** (Matthew 23), while also resisting spiritual laxity. He models fervent devotion without fanaticism, discipline without condemnation.

## Moderation: Islamic Perspective

Moderation (*wasatiyyah*) is explicitly praised in the Qur'an as the mark of a balanced community:

> *"Thus We have made you a justly balanced nation (ummatan wasaṭan)..."* (Qur'an "The Cow" 2:143)

Moderation in Islam manifests as:

- **Avoidance of excess in wealth, consumption, and ritualism**

- **Temperance in judgment,** avoiding harshness or leniency

- **Balanced devotion**, where worldly responsibilities and spiritual obligations complement each other

The Prophet Muhammad said:

> *"Beware of extremism in religion, for those before you were destroyed by it."* (Nasāʾī)

Fasting, charity, family life, and worship in Islam are designed to cultivate rhythmic discipline, not ascetic denial or indulgent excess.

| Criterion | Assessment |
| --- | --- |

| | |
|---|---|
| 1. Communication | ✓ Fosters balanced dialogue and receptivity. |
| 2. Friction | ✓ Avoids **polarizing behavior** and **ideological rigidity**. |
| 3. Problem Solving | ✓ Supports nuanced reasoning and **pragmatic** compromise. |
| 4. Resilience | ✓ Prevents burnout and moral rigidity, sustaining steady effort. |
| 5. Trust & Cooperation | ✓ **Moderates ego** and invites fair collaboration. |
| 6. Adaptability | ✓ Encourages flexibility across changing contexts. |
| 7. Pro-Social w/o Reward | ✓ Resists **attention-seeking** or **moral posturing**. |
| 8. Functional Health | ✓ Linked to reduced stress, better decision-making, and well-being. |

**Moderation: Does it Align with Future Human Growth and Evolution?**

**Moderation: Points of Alignment**

- Both Islam and Christianity treat moderation as a guardrail against moral imbalance and a sign of spiritual maturity.

- Moderation safeguards against zealotry and apathy, producing measured, thoughtful, and effective lives.

- It reflects confidence, not compromise—the strength to remain centered in a turbulent world.

## Moderation: Points of Divergence or Nuance

- Islam more explicitly codifies moderation in ritual, finance, and behavior, particularly through warnings against *ghuluw* (excess).

- Christianity often emphasizes inner restraint and moderation as virtues of sanctified character, especially in the letters of Paul.

- Both traditions confront the danger of fanaticism cloaked in piety, offering moderation as a path of humility and wisdom.

## Moderation Conclusion: Evolutionary Trajectory

Moderation is the spiritual art of equilibrium, a dynamic balance that allows the human spirit to remain upright in a world of extremes. It is not the dullness of compromise but the refinement of discipline, the virtue that sustains all other virtues by tempering excess and restraining deficiency. Islam and Christianity both elevate moderation as a mark of divine wisdom: not a passive neutrality, but a principled calibration of passion, power, and principle. In the evolutionary-spiritual framework, moderation emerges as a cultural immune system, one that preserves coherence amid instability, **resists the polarization that fractures societies**, and **guards freedom from devolving into chaos**. It is the trait that enables complex civilizations to adapt without losing their ethical center. In the future shaped by moral evolution, moderation will be indispensable, **not as restraint for its own sake, but as the strength to flourish in proportion**, to lead without domination, and to progress without implosion.

# Chapter 18: Forgiveness and Reconciliation

**Forgiveness and Reconciliation: Definition and Purpose**

Forgiveness is the moral and spiritual **act of releasing resentment, blame, or the desire for vengeance** toward those who have caused harm. Reconciliation extends forgiveness into restored relationship, rooted in truth, accountability, and renewed trust. Together, they enable the transformation of pain into healing, and alienation into solidarity.

In the evolutionary-spiritual framework, forgiveness and reconciliation are **adaptive strategies for moral repair** and long-term cohesion. They prevent cycles of retaliation, allow reintegration after wrongdoing, and create space for empathy, growth, and lasting peace.

**Forgiveness and Reconciliation: Christian Perspective**

Forgiveness is the heartbeat of the Gospel, modeled by Christ and commanded of His followers.

*"Forgive us our trespasses, as we forgive those who trespass against us."*
(Matthew 6:12)

*"Father, forgive them, for they know not what they do."* (Luke 23:34)

Christian forgiveness is:

- **Unconditional** and proactive, extended even to enemies

- Rooted in divine mercy, not human merit

- Essential for spiritual health and **relational restoration**

Reconciliation is central to Paul's theology:

> *"God was reconciling the world to himself through Christ… and has given us the ministry of reconciliation."* (2 Corinthians 5:18)

**Forgiveness is not forgetting**—it is a creative, redemptive act that opens the future to healing.

**Forgiveness and Reconciliation: Islamic Perspective**

Forgiveness (*ʿafw*) and reconciliation (*ṣulḥ*) are essential themes in Islamic ethics.

The Qur'an teaches:

> *"Let them pardon and overlook. Would you not love for God to forgive you?"* (Qur'an "The Light" 24:22)

God is repeatedly described as Al-Ghafūr (The Forgiving) and Al-ʿAfūw (The Pardoner). Believers are encouraged to:

- Forgive those who wrong them, **especially when in a position of power**

- Reconcile disputes to prevent further harm

- Balance justice with mercy—**seeking fairness but not revenge**

> *"Whoever forgives and makes reconciliation, his reward is with God."* (Qur'an "The Consultation" 42:40)

Islamic law permits retributive justice, but consistently emphasizes forgiveness as a higher moral path, particularly when it leads to peace and moral renewal.

| Criterion | Assessment |
| --- | --- |
| 1. Communication | ✔ **Reopens channels** for dialogue, empathy, and honesty. |
| 2. Friction | ✔ **Halts cycles of violence**, resentment, and retaliation. |
| 3. Problem Solving | ✔ Enables moral **repair**, reintegration, and long-term solutions. |
| 4. Resilience | ✔ Builds emotional strength and societal recovery. |
| 5. Trust & Cooperation | ✔ Restores relationships and institutions after harm. |
| 6. Adaptability | ✔ **Allows for flexible moral response** to complex wrongdoing. |
| 7. Pro-Social w/o Reward | ✔ Often involves moral cost without worldly return. |
| 8. Functional Health | ✔ Linked to lower stress, improved mental health, and social stability. |

**Forgiveness and Reconciliation: Does it Align with Future Human Growth and Evolution?**

**Forgiveness and Reconciliation: Points of Alignment**

- Both Islam and Christianity emphasize that forgiveness is not a sign of weakness but of strength.

- Reconciliation is portrayed as a moral necessity and spiritual opportunity, not merely personal healing.

- Forgiveness is a divine act to be mirrored in human relationships.

## Forgiveness and Reconciliation: Points of Divergence or Nuance

- Islam permits a range of ethical responses to harm: justice, reparation, or forgiveness—with strong spiritual preference for the latter.

- Christianity tends to theologize forgiveness as central to salvation, urging unconditional mercy, as a reflection of God's mercy.

- Islam often grounds reconciliation in community restoration, while Christianity often roots it in communal and personal imitation of Christ.

## Forgiveness and Reconciliation Conclusion: Evolutionary Trajectory

Forgiveness and reconciliation are not mere acts of virtue; they are evolutionary tools for the healing of human history. They transform grievance into grace and rupture into renewal, allowing individuals and communities to emerge wiser from their wounds. Without them, the moral tension of unresolved conflict accumulates until even the strongest societies begin to splinter. With them, past wrongs become raw material for future wisdom. Islam and Christianity both affirm that forgiveness is not a peripheral virtue but a spiritual necessity, the heartbeat of moral maturity and divine emulation. In the evolutionary-spiritual vision, **forgiveness becomes a civilizational strategy**, one that interrupts cycles of revenge, **restores fractured identities**, and expands the human capacity for unity across difference. Those who forgive will not merely soothe the past; **they will forge the future, becoming architects of a more resilient, liberated, and ethically evolved humanity.**

# Chapter 19: Humility

**Humility: Definition and Purpose**

Humility is the clear-eyed **recognition of one's limits**, dependencies, and place within a greater order, paired with a refusal to seek elevation through comparison, dominance, or pride. It does not mean self-negation, but rather self-right-sizing—a moral awareness of both one's value and one's impermanence.

In the evolutionary-spiritual model, humility fosters learning, cooperation, and moral clarity. It is essential for **avoiding the distortions of ego**, building communal trust, and maintaining ethical resilience in the face of success, failure, or power.

**Humility: Christian Perspective**

Humility is foundational in Christian spirituality and moral teaching. Jesus Christ is its supreme exemplar:

> *"He humbled Himself... even to death on a cross."* (Philippians 2:8)

He taught:

> *"For those who exalt themselves will be humbled, and those who humble themselves will be exalted."* (Matthew 23:12)

Christian humility includes:

- Acknowledging God as the source of all goodness

- Serving others without seeking recognition

- Living simply and gratefully, with reverence for life and others

The Beatitudes open with:

> *"Blessed are the poor in spirit, for theirs is the kingdom of heaven."*　(Matthew 5:3)

Humility is the soil in which love, patience, and forgiveness take root.

## Humility: Islamic Perspective

Humility (*tawāḍuʿ*) is among the most praiseworthy traits in Islamic ethics. The Prophet Muhammad is its model, described as approachable, self-effacing, and generous, even while being the leader of a nation.

The Qur'an warns:

> *"Do not walk upon the earth arrogantly. Indeed, you will never split the earth nor reach the mountains in height."* (Qur'an "The Journey by Night" 17:37)

The Prophet said:

> *"Whoever humbles himself for the sake of God, God will raise him."* (Muslim)

Humility in Islam includes:

- Spiritual humility: **acknowledging dependence on God**

- Social humility: respecting others, **regardless of status**

- Intellectual humility: accepting one's limitations and **being open to correction**

Humility is seen as the opposite of arrogance (kibr), which is condemned as the root of Iblīs 'fall and of many societal sins.

| Criterion | Assessment |
|---|---|
| **1. Communication** | ✔ Enhances listening, receptivity, and honest dialogue. |
| **2. Friction** | ✔ Reduces **ego-based conflict** and **moral posturing**. |
| **3. Problem Solving** | ✔ Enables collaborative learning and mutual respect. |
| **4. Resilience** | ✔ Allows graceful response to success, failure, or correction. |
| **5. Trust & Cooperation** | ✔ Fosters credibility, moral trust, and service-oriented leadership. |
| **6. Adaptability** | ✔ Encourages openness to change and complexity. |
| **7. Pro-Social w/o Reward** | ✔ Frees the self from needing recognition to do good. |
| **8. Functional Health** | ✔ Associated with reduced **narcissism** and increased emotional maturity. |

**Humility: Does it Align with Future Human Growth and Evolution?**

**Humility: Points of Alignment**

- Both Islam and Christianity view humility as the **gateway to all other virtues**.

- Humility is not low self-worth but right self-understanding—a necessary posture before God, truth, and others.

- Each tradition affirms that **spiritual arrogance is deadly**, while humility is the path to exaltation.

**Humility: Points of Divergence or Nuance**

- Islam ties humility directly to God-consciousness and justice, condemning pride as not only personal, but socially and theologically dangerous.

- Christianity often emphasizes Christ's humility as redemptive and urges believers to imitate that self-emptying but God glorifying love.

- Islam discourages public displays of superiority in wealth or knowledge; Christianity similarly exalts the meek and the servant-hearted.

**Humility Conclusion: Evolutionary Trajectory**

Humility is the root system of moral and evolutionary growth. It creates the internal space where truth can be received, wisdom can take root, and love can be offered without performance or pride. Both Islam and Christianity place humility at the heart of the spiritual life, not as weakness, but as a reflection of divine majesty and a gateway to deeper self-awareness. In the evolutionary-spiritual framework, humility is not just a personal virtue; it is a civilizational necessity. It tempers the arrogance that leads to conflict, the tribalism that fractures humanity, and the hubris that hastens ecological and moral collapse. Humility **enables adaptive learning**, mutual respect, and ethical innovation, traits essential for long-term survival in an interconnected world. **In the evolutionary**

**future, it will be humility, rather than dominance, that marks the truly advanced**: those capable of reverence over conquest, service over self-glory, and restraint over recklessness.

# Sacred Evolution

# Part II: Societal Ethics and Vices

# Chapter 20: Marriage and Family

**Marriage and Family: Definition and Purpose**

Marriage and family are foundational social institutions that organize intimacy, reproduction, caregiving, and moral education. They are not merely private relationships but cultural ecosystems that shape intergenerational transmission of values, emotional resilience, and community stability.

In the evolutionary-spiritual model, marriage and family serve as frameworks of mutual responsibility, **enabling the formation of cooperative bonds that extend beyond biology**. A morally healthy society depends on the strength, justice, and compassion embedded in its families.

**Christian Perspective**

Marriage is a covenant relationship in Christian theology, reflecting both human intimacy and divine mystery:

> *"For this reason a man shall leave his father and mother… and the two shall become one flesh."* (Matthew 19:5)

Paul likens marriage to the union between Christ and the Church (Ephesians 5:25–33), emphasizing:

- Mutual self-giving

- Sacrificial love

- Lifelong fidelity and unity

Family is described as the domestic church—a setting for:

- Nurturing children in love and moral wisdom

- Practicing forgiveness, service, and communal worship

- Passing on faith through relational example and instruction

While models of family differ across Christian traditions, all affirm the sacredness of marriage and the centrality of family as a moral training ground.

**Marriage and Family: Islamic Perspective**

Marriage (*nikāḥ*) is a sacred contract in Islam, **considered half of faith for those capable**. The Qur'an describes it as a sign of divine love and balance:

> *"And among His signs is that He created for you spouses from among yourselves, that you may find tranquility in them, and He placed between you affection and mercy."* (Qur'an "The Romans" 30:21)

Family is central to Islamic life:

- Mutual **rights and responsibilities** between spouses

- Emphasis on parental care, filial respect, and **generational continuity**

- Recognition of the extended family as **part of the social and spiritual network**

The Prophet Muhammad's own family life exemplified:

- Gentleness, consultation, and loyalty

- Raising children with moral grounding and affection

- Maintaining **fairness and balance within the household**

Islamic law provides detailed guidance on marriage contracts, inheritance, and family rights to ensure justice and compassion at the heart of the home.

| Criterion | Assessment |
|---|---|
| **1. Communication** | ✔ Builds trust-based environments for safe emotional expression. |
| **2. Friction** | ✔ Structures relational boundaries and roles, reducing chaos. |
| **3. Problem Solving** | ✔ Offers generational wisdom, conflict resolution, and shared responsibility. |
| **4. Resilience** | ✔ Strengthens **emotional and financial support networks**. |
| **5. Trust & Cooperation** | ✔ Teaches loyalty, compromise, and long-term collaboration. |
| **6. Adaptability** | ✔ Core values remain stable even as family structures evolve culturally. |
| **7. Pro-Social w/o Reward** | ✔ Marriage and parenthood demand sacrificial love without immediate gain. |

| 8. Functional Health | ✔ Correlated with reduced crime, better child outcomes, and well-being. |
|---|---|

## Marriage and Family: Does it Align with Future Human Growth and Evolution?

### Marriage and Family: Points of Alignment

- Both Islam and Christianity view marriage as a divinely sanctioned institution rooted in responsibility, fidelity, and mutual support.

- **Family is more than biology**—it is a moral community where love and duty meet.

- Both traditions uphold parenting as a sacred task and marriage as a moral covenant.

### Marriage and Family: Points of Divergence or Nuance

- Islam includes polygyny under regulated conditions, though monogamy is normative in practice and encouraged in many contexts.

- Christianity almost universally supports monogamy, often as a symbol of spiritual exclusivity and union.

- Islamic ethics legally structure marriage and family obligations; Christian ethics emphasize sacramental grace and spiritual mutuality, especially in marital love.

### Marriage and Family Conclusion: Evolutionary Trajectory

Marriage and family are not outmoded relics; they are dynamic ecosystems of love, sacrifice, and resilience that anchor human development across generations. Both Islam and Christianity affirm that no society can flourish without families rooted in care, sustained by fidelity, and infused with mercy. In the evolutionary-spiritual framework, marriage and family are more than social contracts; **they are moral incubators and evolutionary sanctuaries where character is shaped, empathy is cultivated, and future citizens are formed in the rhythms of responsibility and belonging.** They are the first schools of trust, where patience is practiced, forgiveness is modeled, and love is made durable through trial and grace. In a future shaped by ethical evolution, marriage and family will remain the vital ground where humanity learns to balance freedom with commitment and individuality with interdependence. They are not merely structures of shelter; they are crucibles in which souls are tempered, values transmitted, and civilization reborn with every generation.

# Chapter 21: Education

**Education: Definition and Purpose**

Education is the structured process by which knowledge, skills, values, and moral frameworks are transmitted across individuals and generations. It is not limited to formal schooling—it includes spiritual formation, ethical reasoning, critical thinking, and the cultivation of wisdom.

In the evolutionary-spiritual framework, **education is a strategic lever of civilizational resilience**. It expands human potential, fosters adaptability, and embeds virtues into the very consciousness of a society.

**Education: Christian Perspective**

Christianity views education as both intellectual and spiritual formation. The Hebrew tradition emphasized teaching children diligently (Deuteronomy 6:7), and Jesus was called "Teacher" more than any other title in the Gospels.

*"The fear of the Lord is the beginning of wisdom."* (Proverbs 9:10)

Christian education includes:

- Scriptural literacy: reading, interpreting, and embodying sacred texts

- Moral formation: cultivating virtues like humility, service, and compassion

- **Liberal learning**: integrating truth, beauty, and goodness through philosophy, the arts, and science

The early Church established schools, monasteries, and universities, often serving as the preservers and transmitters of civilization through dark ages and upheaval.

Paul instructs:

> *"Be transformed by the renewing of your mind."* (Romans 12:2)

**Education: Islamic Perspective**

Education is **sacred** in Islam. The first Qur'anic revelation— *"Read!"* *(Iqra)*—sets the tone for a religion built on learning, literacy, and reflection.

> *"Are those who know equal to those who do not know?"* (Qur'an 39:9)

The Prophet Muhammad said:

> *"Seeking knowledge is an obligation upon every Muslim, male and female."* (Ibn Mājah)

Islamic education includes:

- Revelation-based knowledge (ʿilm al-naqlī): Qur'an, Hadith, theology

- Empirical and rational sciences (ʿilm al-ʿaqlī): medicine, astronomy, philosophy

- Moral and spiritual disciplines: tazkiyah (soul purification), adab (ethical comportment)

Historically, Islamic civilization produced scholars, scientists, and ethicists in global centers of learning like Baghdad, Cordoba, and Cairo—demonstrating that education is central to faith and flourishing.

| Criterion | Assessment |
|---|---|
| 1. Communication | ✔ Develops language, empathy, and cross-cultural understanding. |
| 2. Friction | ✔ Reduces ignorance, prejudice, and reactive conflict. |
| 3. Problem Solving | ✔ **Enables innovation**, moral reasoning, and scientific progress. |
| 4. Resilience | ✔ Equips individuals and communities to **adapt and rebuild**. |
| 5. Trust & Cooperation | ✔ Shared knowledge promotes **civic engagement and solidarity**. |
| 6. Adaptability | ✔ Lifelong learning supports **growth across eras and crises**. |
| 7. Pro-Social w/o Reward | ✔ Knowledge often pursued for truth's sake, not profit. |

| | |
|---|---|
| **8. Functional Health** | ✓ Correlated with improved well-being, social mobility, and community development. |

## Education: Does it Align with Future Human Growth and Evolution?

### Education: Points of Alignment

- Both Islam and Christianity treat education as a spiritual and civic duty, not a luxury.

- They affirm that the pursuit of knowledge is a sacred act, essential for justice, dignity, and faith.

- Each tradition has historic institutions dedicated to teaching that combine reason with revelation.

### Education: Points of Divergence or Nuance

- Islam maintains a strong linkage between religious and worldly knowledge, viewing both as expressions of God's signs (*āyāt*).

- Christianity historically distinguished more between secular and sacred learning, though modern integration is increasingly common.

- Islamic ethics see learning as part of worship and responsibility, while Christian ethics frame it as transformation and service.

### Education Conclusion: Evolutionary Trajectory

Education is the evolutionary bridge from instinct to insight, from inherited dogma to generative wisdom. It is the mechanism by which civilizations renew their moral compass, recalibrate their collective memory, and prepare the next generation not just to survive, but to evolve. Islam and Christianity both uphold a vision of education rooted in reverence for the sacred, expanded by disciplined inquiry, and directed toward justice, compassion, and transcendence. In the evolutionary-spiritual framework, **education becomes humanity's most vital tool for adaptive advancement**. It cultivates self-awareness, ethical reasoning, and a shared language of dignity, enabling diverse communities to navigate complexity with conscience rather than coercion. Education is not merely preparation for life; it is the sacred means by which life becomes meaningful, culture becomes conscious, and evolution becomes ethical.

# Chapter 22: Nationalism

## Nationalism: Definition and Purpose

Nationalism is the ideological commitment to political, cultural, and emotional allegiance to a nation, often characterized by shared language, ethnicity, history, or values. While it can promote unity and civic pride, nationalism also carries the risk of exclusivism, conflict, and moral myopia if elevated above ethical principles or universal compassion.

In the evolutionary-spiritual framework, nationalism must be examined for its potential to promote group cohesion without devolving into tribalism, authoritarianism, or dehumanization. A spiritually mature nationalism is one that serves justice and human dignity, not one that subordinates moral truth to group identity.

## Nationalism: Christian Perspective

Christianity views ultimate belonging as citizenship in the Kingdom of God (Philippians 3:20). Early Christians were often stateless, persecuted minorities who emphasized spiritual unity over political identity.

Jesus teaches:

> *"Love your neighbor as yourself."* (Matthew 22:39)

And the parable of the Good Samaritan extends that neighborliness across cultural and ethnic lines.

Yet historically, Christianity has also been intertwined with national empires, state churches, and colonial expansion—often distorting faith into national ideology.

Contemporary Christian ethics challenge believers to:

- Love the nation, **but not worship it**

- Be loyal citizens, but also prophetic voices for justice

- Embrace national responsibility while **maintaining global compassion**

**Nationalism: Islamic Perspective**

Islam promotes ummah consciousness—a sense of spiritual solidarity that transcends ethnic and national boundaries. While not inherently opposed to national identity, Islamic ethics warn strongly against asabiyyah (tribal or ethnic chauvinism):

> *"He is not one of us who calls for asabiyyah."* (Abu Dawud)

The Qur'an teaches:

> *"O mankind, We created you from male and female and made you peoples and tribes that you may know one another…"* (Qur'an "The Chambers" 49:13)

Islamic thought acknowledges:

- The legitimacy of nation-states as structures for governance and law

- The ethical need to **resist exclusionary, supremacist, or unjust nationalism**

- That moral allegiance lies ultimately with God, **not with flag, ethnicity, or race**

Historically, Islam produced both multiethnic empires and pan-Islamic movements, reflecting a vision of moral unity over national division.

| Criterion | Assessment |
|---|---|
| **1. Communication** | ! Can unite in-group communication, but **hinder dialogue** across boundaries. |
| **2. Friction** | ! May reduce internal chaos, but can **exacerbate external conflict.** |
| **3. Problem Solving** | ✓ Supports organized governance when tempered by ethics. |
| **4. Resilience** | ! Builds strong identity under threat, but may **reduce moral flexibility.** |
| **5. Trust & Cooperation** | ! Works in-group; may erode cross-cultural trust without moral guardrails. |
| **6. Adaptability** | ! Often rigid; can resist change or pluralism. |
| **7. Pro-Social w/o Reward** | ! Altruism may be conditional on group membership. |
| **8. Functional Health** | ! National pride can inspire responsibility—or fuel division. |

**Nationalism: Does it Align with Future Human Growth and Evolution?**

**Nationalism: Points of Alignment**

- Both Islam and Christianity acknowledge that cultural identity and political belonging can be meaningful and ethical.

- Neither tradition accepts unquestioning loyalty to unjust authority or exclusionary nationalism.

- Both call for universal moral frameworks to guide national vision—rooted in justice, compassion, and accountability to God.

**Nationalism: Points of Divergence or Nuance**

- Islam elevates religious brotherhood (ummah) as the highest allegiance, which can sometimes challenge nationalist constructs.

- Christianity often embraces dual citizenship—with allegiance to both the state and God, but insists that when they conflict, conscience must obey the higher law.

- Islam incorporates legal critique of nationalism, while Christianity often functions through prophetic resistance and ethical witness.

**Nationalism Conclusion: Evolutionary Trajectory**

Nationalism is an ethically **double-edged force**, capable of protecting identity or provoking division, fostering solidarity or fueling supremacy. Both Islam and Christianity challenge us to transcend attachments based solely on blood, soil, or border, and to root our loyalties in justice, mercy, and humility before a higher moral order. In the evolutionary-spiritual framework, nationalism is not inherently regressive, but it must evolve. If shaped by ethical restraint and spiritual vision, it can mature into responsible civic stewardship. If left unchecked, it risks **devolving into tribalism, fear, and spiritual contraction**. A morally advanced civilization does not demand the abandonment of national belonging; rather, it demands that such belonging be held in tension with universal truth, shared dignity, and the well-being of all creation. In the evolutionary future, the greatest love of homeland will be measured not by exclusion or pride, but by how it teaches its people to **serve humanity beyond its borders and to love justice more than the flag.**

# Chapter 23: The Seven Deadly Sins

**Seven Deadly Sins: Definition and Purpose**

The Seven Deadly Sins are classical categories of habitual moral failure that undermine character, distort desire, and fragment community. They are not mere mistakes, but deep-seated inclinations that corrupt the soul from within. The seven traditionally named are: pride, greed, lust, envy, gluttony, wrath, and sloth.

In the evolutionary-spiritual framework, these vices are understood not simply as theological transgressions, but as maladaptive behavioral patterns that disrupt trust, destabilize relationships, and weaken communal integrity and spiritual vitality.

**Seven Deadly Sins: Christian Perspective**

The Seven Deadly Sins emerged as a moral taxonomy in early Christian monasticism (Evagrius, Gregory the Great), serving as a diagnostic tool for spiritual self-examination.

Each vice is a perversion of a virtue:

- Pride vs. humility

- Greed vs. charity

- Lust vs. chastity

- Envy vs. gratitude

- Gluttony vs. temperance

- Wrath vs. patience

- Sloth vs. diligence

Paul warns against such behaviors in multiple epistles:

*"The acts of the flesh are obvious: sexual immorality, impurity… hatred, discord, jealousy, fits of rage…"* (Galatians 5:19–21)

Christian tradition emphasizes:

- Grace-driven transformation

- Daily confession and renewal

- Spiritual disciplines to uproot vice and grow virtue

**Seven Deadly Sins: Islamic Perspective**

Islam does not organize moral failings under the exact rubric of "seven deadly sins," but it identifies major sins (kab'āir) and diseases of the heart (amrāḍ al-qalb) that closely correspond to these vices.

Examples include:

- Pride (kibr): condemned as the sin of Iblīs, who refused to bow to Adam

- Greed (ḥirs) and envy (ḥasad): destructive forces condemned in the Qur'an

- Lust (shahwah): subject to regulation and spiritual discipline

- Anger (ghaḍab) and sloth (kasl): obstacles to ethical and spiritual maturity

- Gluttony (isrāf): overconsumption is forbidden, as God does not love the extravagant

The Prophet Muhammad warned:

> *"There is a piece of flesh in the body: if it is sound, the whole body is sound. If it is corrupted, the whole body is corrupted. It is the heart."*
> (Bukhari & Muslim)

Islam encourages purification through:

- Regular worship

- Charity and self-restraint

- Repentance and spiritual awareness

| Criterion | Assessment |
|---|---|
| **1. Communication** | ! These sins distort empathy, dialogue, and emotional honesty. |
| **2. Friction** | ! Left unchecked, they fuel jealousy, conflict, exploitation, and isolation. |
| **3. Problem Solving** | ! Vices cloud judgment, breed self-deception, and sabotage collaboration. |
| **4. Resilience** | ! These sins erode willpower, discipline, and long-term coherence. |

| | |
|---|---|
| **5. Trust & Cooperation** | ! Undermine moral trust and social reliability. |
| **6. Adaptability** | ✓ Identifying and correcting these vices supports personal evolution. |
| **7. Pro-Social w/o Reward** | ✓ Resisting these urges reflects internalized moral maturity. |
| **8. Functional Health** | ! Strong correlation between unchecked vice and addiction, violence, or alienation. |

**Seven Deadly Sins: Does it Align with Future Human Growth and Evolution?**

**Seven Deadly Sins: Points of Alignment**

- Both Islam and Christianity view these moral failings as spiritually corrosive and socially destructive.

- They emphasize that these vices are root-level distortions that require ongoing discipline, repentance, and transformation.

- Both offer concrete practicespr: prayer, fasting, charity, reflection, to counter and heal these tendencies.

**Seven Deadly Sins: Points of Divergence or Nuance**

- Christianity formalized these vices into a spiritual inventory, used for personal confession and pastoral care.

- Islam integrates the concept of moral disease into a broader framework of legal accountability and inner purification, without fixing the list to a rigid set.

- Both agree that these are not merely private flaws, but community-level threats to moral ecology.

## Seven Deadly Sins Conclusion: Evolutionary Trajectory

The Seven Deadly Sins are not relics of medieval theology; they are enduring signals of inner disorder, evolutionary missteps that threaten both individual flourishing and societal stability. Pride and greed destabilize communities and corrupt leadership; lust and wrath fracture families and distort intimacy; envy corrodes gratitude and social harmony; gluttony and sloth erode discipline and dull the moral will. Both Islam and Christianity offer more than condemnation; they offer pathways of inner transformation, where vice is neither indulged nor suppressed, but refined into its higher counterpart. Within the evolutionary-spiritual framework, the struggle against these distortions is not an act of repression, but of conscious liberation, an essential step in evolving beyond the reactive mind and toward the integrated self. As humanity matures, **the disciplines that diminish these vices will be recognized not as moral restraints, but as the architecture of resilience and ethical vitality**. In the future we must build, healing the world begins with healing character.

# Chapter 24: Exploitation and Addictive Systems

**Exploitation and Addictive Systems: Definition and Purpose**

Exploitation refers to systems or behaviors that take unfair advantage of others 'vulnerability, while addictive systems refer to structures that create dependency, compulsion, and moral erosion, whether physiological, financial, or psychological. Together, they form patterns of ethical breakdown that appear legal or normal but undermine individual dignity and communal well-being.

In the evolutionary-spiritual framework, such systems are understood as **pathological distortions of trust**, autonomy, and purpose. They weaken moral agency, foster inequality, and fragment societies from within, even when dressed in economic or social legitimacy.

**Exploitation and Addictive Systems: Christian Perspective**

Christian ethics also critique **systems of bondage**—whether to wealth, substances, or impulses—that compromise freedom and virtue.

Jesus warns:

> *"You cannot serve both God and money."* (Matthew 6:24)

Paul teaches:

> *"All things are lawful for me, but I will not be dominated by anything."* (1 Corinthians 6:12)

Key concerns:

- Exploitation of the poor through **unjust labor or predatory lending**

- Addictions that degrade the body, the "temple of the Holy Spirit" (1 Corinthians 6:19)

- Spiritual captivity: when habits or systems enslave moral agency

Christian theology frames liberation not just in political terms, but as freedom from sin, ego, and compulsion, made possible by:

- Grace and repentance

- Community accountability

- Embodied virtue and wise stewardship

**Exploitation and Addictive Systems: Islamic Perspective**

Islamic ethics are deeply concerned with the exploitation of others through unjust gain or moral corruption. Two prominent examples are:

**a) Usury (Ribā)**

> *"God has permitted trade and forbidden usury."* (Qur'an "The Cow" 2:275)

Usury is condemned as a system that:

- Extracts wealth without effort or risk

- Preys on the poor and indebted

- **Distorts economic justice**

Islam permits profit through risk and productivity, but forbids guaranteed profit from debt, encouraging partnership-based finance (e.g., *mushārakah, murābaḥah*).

## b) Intoxicants and Gambling

> *"Intoxicants, gambling... are abominations of Satan's handiwork. Avoid them so that you may succeed."* (Qur'an 5:90)

These are viewed as:

- **Moral threats** that erode judgment and self-control

- Social cancers linked to violence, neglect, and poverty

- Spiritual veils that distance the soul from God and purpose

Islam encourages detoxification of the self and society through:

- Fasting, prayer, and charity

- Community support and ethical alternatives

- Systemic reform, not just individual discipline

| Criterion | Assessment |
|---|---|
| **1. Communication** | ! Exploitative systems distort motives and social contracts. |
| **2. Friction** | ! They increase inequality, resentment, and destabilization. |

| | |
|---|---|
| **3. Problem Solving** | ! Addictive loops impair clarity and accountability. |
| **4. Resilience** | ! These systems erode personal resilience and social safety nets. |
| **5. Trust & Cooperation** | ! Exploitation breaks trust; addiction isolates. |
| **6. Adaptability** | ✓ Ethical reform and recovery models foster long-term adaptability. |
| **7. Pro-Social w/o Reward** | ✓ Acts of restraint and reform require sacrifice and moral strength. |
| **8. Functional Health** | ! Exploitation and addiction correlate with health crises, family breakdown, and despair. |

**Exploitation and Addiction: Does it Align with Future Human Growth and Evolution?**

**Exploitation and Addictive Systems: Points of Alignment**

- Both Islam and Christianity condemn systems that exploit human weakness for profit, pleasure, or control.

- Both affirm that freedom is not indulgence, but moral clarity and ethical self-mastery.

- The path of liberation includes spiritual reform, personal discipline, and communal accountability.

**Exploitation and Addictive Systems: Points of Divergence or Nuance**

- Islam legally prohibits usury and intoxicants outright, offering systemic structures for ethical finance and moral behavior.

- Christianity varies by tradition—some emphasize temperance over prohibition and often rely on pastoral care and community support rather than law.

- Both traditions recognize that spiritual freedom requires structural transformation, not just private virtue.

**Exploitation and Addictive Systems Conclusion: Evolutionary Trajectory**

Exploitation and addiction represent evolutionary regressions that distort genuine autonomy and hinder the ethical advancement envisioned by the evolutionary-spiritual framework. Rather than enhancing freedom, these patterns **trap humanity in cycles of dependency**, fragmentation, and systemic decline. Their persistence disrupts cooperative potential, undermines communal trust, and erodes moral clarity, qualities essential for sustainable evolution. Islam and Christianity strongly align with this framework, as both traditions urge humanity to reclaim the inner dignity and collective justice that exploitation and addiction obscure. They advocate for a vision of freedom grounded not in the absence of boundaries but in the disciplined refinement of desire, thus supporting a coherent trajectory toward greater communal harmony and individual integrity. In this evolutionary-spiritual context, **true freedom emerges as the disciplined capacity to choose dignity overindulgence, purpose over compulsion**, and the sacred over the superficial. By embracing this ethical alignment, humanity can evolve beyond predatory systems and addictive cycles, cultivating instead **resilient cultures built on restraint, reverence, and shared purpose.**

# Chapter 25: Protection of the Vulnerable and Marginalized

**Protection of the Vulnerable and Marginalized: Definition and Purpose**

Protecting the vulnerable and marginalized involves safeguarding those who lack power or protection, including the poor, orphans, widows, refugees, disabled, elderly, and socially ostracized. This virtue transforms ethical theory into action by ensuring that compassion and justice are extended beyond privilege.

In the evolutionary-spiritual framework, **protection of the vulnerable is a litmus test of moral maturity**. Societies that care for their weakest members build resilience, trust, and sacred credibility, while those that ignore them invite fragmentation, revolt, or spiritual decay.

**Protection of the Vulnerable and Marginalized: Christian Perspective**

The Bible, especially the Old Testament prophets and Jesus, consistently center the marginalized as sacred subjects of divine concern:

> *"Learn to do right; seek justice. Defend the oppressed. Take up the cause of the fatherless; plead the case of the widow."* (Isaiah 1:17)

Jesus identifies with the vulnerable:

> *"Whatever you did for the least of these… you did for me."* (Matthew 25:40)

Key Christian emphases include:

- Almsgiving, hospitality, and structural advocacy

- Valuing the sick, poor, and outcast as bearers of divine presence

- The early Church's provision for widows and orphans (Acts 6)

In Christian ethics, moral authority often rests with those on the margins, not those at the center of power.

## Protection of the Vulnerable and Marginalized: Islamic Perspective

The Qur'an repeatedly ties faith to the treatment of the vulnerable:

*"Have you seen the one who denies the religion? It is he who repels the orphan and does not encourage the feeding of the poor."* (Qur'an "Small Kindnesses" 107:1–3)

Key protected groups include:

- Orphans (*yatāmā*): given legal, financial, and spiritual priority

- Widows: ensured inheritance and protection from neglect

- The poor, travelers, captives, and neighbors: mentioned in zakāt and social responsibilities

The Prophet Muhammad was an orphan himself and said:

*"I and the one who cares for an orphan will be together in Paradise like this,"* and he held two fingers side by side. (Bukhari)

Islamic systems include:

- Zakāt, Khums and ṣadaqah as redistributive tools

- Legal guardianship and fiduciary duties

- Emphasis on equitable dignity, regardless of status

| Criterion | Assessment |
|---|---|
| **1. Communication** | ✔ Validates unheard voices and affirms shared humanity. |
| **2. Friction** | ✔ Prevents resentment, social unrest, and abandonment. |
| **3. Problem Solving** | ✔ Includes diverse experiences and expands moral imagination. |
| **4. Resilience** | ✔ Strengthens social bonds and crisis response mechanisms. |
| **5. Trust & Cooperation** | ✔ Fosters inclusive loyalty and moral solidarity. |
| **6. Adaptability** | ✔ Demonstrates ethical flexibility across diverse needs and populations. |
| **7. Pro-Social w/o Reward** | ✔ Caring for the vulnerable is often done without recognition. |
| **8. Functional Health** | ✔ Correlates with stronger institutions, better public health, and lower violence. |

**Protection of Vulnerable: Does it Align with Future Human Growth and Evolution?**

**Protection of the Vulnerable and Marginalized: Points of Alignment**

- Both Islam and Christianity place care for the vulnerable at the heart of religious faithfulness.

- The presence of vulnerable persons is seen not as a burden, but as a test and refinement of communal virtue.

- Acts of care are framed not only as social service, but as sacred transactions with God.

**Protection of the Vulnerable and Marginalized: Points of Divergence or Nuance**

- Islam integrates protection through legal structures and economic obligations, such as zakāt allocation and guardian responsibilities.

- Christianity emphasizes development of both personal charity as well as institutional based systems to care for the marginalized.

- Both are evolving to respond to new categories of vulnerability: displaced persons, mentally ill, LGBTQ individuals, and climate refugees.

**Protection of the Vulnerable and Marginalized Conclusion: Evolutionary Trajectory**

A society's spiritual and evolutionary credibility is measured by how it treats its most vulnerable. From the orphan and the widow to the outcast and the stranger, those on the margins reveal the moral architecture of a civilization. Islam and Christianity both teach that neglecting the weak is not just a lapse in compassion; it is a rupture in the covenant of justice.

Within the evolutionary-spiritual framework, protecting the vulnerable is not an act of charity, but a strategy of resilience. It fosters trust, strengthens communal bonds, and ensures that no segment of the human family is left to fall into despair or exploitation. **In the future shaped by ethical evolution, progress will no longer be defined by power, wealth, or technological advancement, but by our willingness to dignify the overlooked and uplift the excluded**. The true strength of a society will be found not in how high it rises, but in how deeply it roots itself in the moral care of those most easily ignored.

# Chapter 26: War and Peace

## War and Peace: Definition and Purpose

War is the **organized use of violence** to achieve political or ideological ends; peace is the state of justice, mutual respect, and nonviolence among individuals, communities, or nations. In moral traditions, the question is not simply whether war exists, but under what conditions it is justified, and more importantly, how peace can be created, protected, and sustained.

In the evolutionary-spiritual framework, **peace is not just the absence of conflict but the presence of moral architecture**: systems of justice, empathy, forgiveness, and shared accountability. War may sometimes be unavoidable, but peace is always preferable, transformative, and evolutionary.

## War and Peace: Christian Perspective

Jesus promotes a vision of radical peacemaking and nonviolence:

> *"Blessed are the peacemakers, for they shall be called children of God."*
> (Matthew 5:9)

> *"Put your sword back in its place..."* (Matthew 26:52)

However, Christian theology developed a Just War Theory (Augustine, Aquinas), outlining when war might be morally permissible:

- **Just** cause (defense, protection of the innocent)

- Right **intention**

- **Last resort**

- **Proportional** response

- Discrimination between combatants and civilians

Today, many Christian denominations embrace peacebuilding, restorative justice, and interfaith diplomacy, while others maintain traditional doctrines of moral resistance through force when necessary.

**War and Peace: Islamic Perspective**

Islam permits war under tightly restricted ethical guidelines, primarily for:

- Self-defense

- Protection of religious freedom

- Response to oppression

*"Fight in the way of God those who fight you, but do not transgress. Indeed, God does not love transgressors."* (Qur'an "The Cow" 2:190)

Key ethical principles in Islamic just war theory:

- **No harm to non-combatants** (women, children, elderly)

- Protection of **nature, property, and prisoners**

- Pursuit of peace when possible:

*"And if they incline to peace, then incline to it also"*
(Qur'an "The Bounties" 8:61)

The Prophet Muhammad's military actions were governed by strict codes, often preferring treaties, mercy, and reconciliation over force. Jihad, often misunderstood, primarily refers to moral struggle, not violence.

| Criterion | Assessment |
|---|---|
| **1. Communication** | ✔ Peace fosters honest dialogue and sustained trust. |
| **2. Friction** | ✔ Peace reduces cycles of trauma and retaliation. |
| **3. Problem Solving** | ✔ Nonviolent systems resolve disputes with moral and practical wisdom. |
| **4. Resilience** | ✔ Peaceful societies are more adaptive and sustainable. |
| **5. Trust & Cooperation** | ✔ Just peace promotes deeper social and global cooperation. |
| **6. Adaptability** | ✔ Peace-centered cultures navigate diversity without fragmentation. |
| **7. Pro-Social w/o Reward** | ✔ Peacemaking often involves sacrifice without material benefit. |

| | |
|---|---|
| **8. Functional Health** | ✔ War correlates with trauma and collapse; peace with flourishing. |

## War and Peace: Does it Align with Future Human Growth and Evolution?

### War and Peace: Points of Alignment

- Both Islam and Christianity uphold peace as the preferred moral and spiritual state.

- Both traditions allow for war only under **stringent ethical guidelines**, with a call to limit harm and restore justice.

- Peacemaking is not viewed as weakness but as moral strength and divine alignment.

### War and Peace: Points of Divergence or Nuance

- Islam allows defensive warfare through juridical ethics, balancing force with compassion, and linking it to communal integrity.

- Christianity includes stronger pacifist strands, especially in the teachings of Jesus and early Christian martyrs.

- Both traditions support peacebuilding efforts today, including conflict resolution, reconciliation, and global justice advocacy.

### War and Peace Conclusion: Evolutionary Trajectory

Peace is not naïve idealism; it is an **evolutionary imperative**. While war may have shaped the rise and fall of past civilizations, it now threatens the very survival of an interconnected species. Islam and Christianity both call us to build moral orders grounded not in tribal loyalty, but in universal justice and sacred restraint. In the evolutionary-spiritual framework, war is increasingly understood as **a symptom of unresolved moral trauma and structural failure**, a **regressive reflex** from an earlier stage of collective development. Peace, by contrast, is not the absence of conflict but the presence of ethical design: systems guided by empathy, justice, and a shared vision of human flourishing. In the future, those who choose peace will not be seen as passive; **they will be the architects of a more conscious civilization**, where power serves principle and where justice flows not from conquest but from character, compassion, and moral evolution.

# Chapter 27: Care of Animals

## Care of Animals: Definition and Purpose

Care of animals refers to the ethical treatment, protection, and compassionate stewardship of non-human creatures. It is grounded in the recognition that animals are living beings with needs, emotions, and divine purpose, not merely resources for human use. Humane treatment of animals reflects a society's moral refinement and spiritual consciousness.

In the evolutionary-spiritual framework, how we treat animals shapes our capacity for empathy, restraint, and justice. Respect for life beyond the human realm cultivates a reverence for creation, prevents cruelty from becoming systemic, and models ethical stewardship over domination.

## Care of Animals: Christian Perspective

Christianity affirms that all creatures are part of God's good creation:

*"God saw all that He had made, and it was very good."* (Genesis 1:31)

Humanity is given dominion, but not domination:

*"The righteous care for the needs of their animals."* (Proverbs 12:10)

Christian tradition, especially through figures like St. Francis of Assisi, views animals as:

- Co-creatures, reflecting the Creator's wisdom

- Symbols of innocence, beauty, and divine presence

- Recipients of care, not exploitation

Many Christian theologians and movements today advocate for:

- **Ethical farming** and food systems

- Protection of **endangered species**

- A theology of creation that includes animal well-being

**Care of Animals: Islamic Perspective**

Islam views animals as communities with purpose, not as property or tools:

> *"There is no creature on earth or bird that flies with its wings but they are communities like you."* (Qur'an "The Cattle" 6:38)

Key principles include:

- Mercy in treatment: even in slaughter, Islam demands the sharpest blade, the least suffering

- Prohibition of cruelty: harming animals without cause is a grave sin

- Rights of animals: to food, rest, and protection from abuse

The Prophet Muhammad exemplified care for animals:

- Forbade overburdening camels and donkeys

- Described divine forgiveness for one who gave water to a thirsty dog

- Described hellfire for one who starved a cat

Animals, in Islam, are seen as signs of God, and moral treatment of them is linked to spiritual reward or punishment.

| Criterion | Assessment |
|---|---|
| 1. Communication | ✓ Expands empathy and reverence across species boundaries. |
| 2. Friction | ✓ Reduces systemic violence and emotional desensitization. |
| 3. Problem Solving | ✓ Encourages sustainable food systems and ecological ethics. |
| 4. Resilience | ✓ Cultivates discipline, compassion, and balance in human-animal interactions. |
| 5. Trust & Cooperation | ✓ Models stewardship and responsibility in interspecies relationships. |
| 6. Adaptability | ✓ Aligns with global shifts toward sustainability and animal welfare. |
| 7. Pro-Social w/o Reward | ✓ Humane treatment of animals often lacks material gain but builds moral capital. |
| 8. Functional Health | ✓ Ethical care improves ecosystems, human health, and emotional development. |

**Care of Animals: Does it Align with Future Human Growth and Evolution?**

## Care of Animals: Points of Alignment

- Both Islam and Christianity affirm that animal life is sacred, and that **cruelty toward animals is a spiritual failing.**

- Ethical treatment of animals is not peripheral—it is a **moral mirror of how we treat the weak, the silent, and the dependent.**

- Each tradition upholds the role of the human being as steward, not exploiter, of creation.

## Care of Animals: Points of Divergence or Nuance

- Islam provides specific legal regulations for the care and slaughter of animals, deeply embedded in daily practice (e.g., halal guidelines).

- Christianity more often promotes relational and symbolic ethics, with emphasis on gentle dominion and reverence for creation.

- Islam encourages practical respect; Christianity often recognizes the bond as part of God's unfolding revelation in nature.

## Care of Animals Conclusion: Evolutionary Trajectory

The way we treat animals is a mirror of our evolutionary maturity. Cruelty calcifies the soul and traps us in patterns of dominance and detachment; compassion softens the heart and opens the path to deeper interdependence. Both Islam and Christianity teach that dominion over creation is not a license to exploit, but a sacred test of mercy, humility, and stewardship. In the evolutionary-spiritual framework, ethical care of animals is not sentimentalism; **it is a civilizational marker of moral development.** It reflects the transition from survival through control to flourishing through empathy. As humanity evolves, our relationship with animals will be redefined, not as masters over the voiceless, but as guardians within a shared web of life. In this future, the strong will be measured not by how much they take, but by how gently they protect, and compassion will begin not at the top of the hierarchy, but at its most vulnerable edge.

# Chapter 28: Environmental Stewardship

### Environmental Stewardship: Definition and Purpose

Environmental stewardship is the moral and spiritual responsibility to protect, sustain, and honor the natural world. It entails recognizing that the earth is not human property but a sacred trust, and that humans are called to be caretakers, not consumers.

In the evolutionary-spiritual framework, environmental stewardship reflects maturity in our relationship with creation. It fosters long-term survival, preserves biodiversity, and signals a civilization that chooses sustainability over exploitation, and cooperation with nature over dominance of it.

### Environmental Stewardship: Christian Perspective

Christianity begins with the creation story, in which God sees the world as "very good" and entrusts it to humanity's care:

> *"The Lord God took the man and put him in the garden... to work it and take care of it."* (Genesis 2:15)

**Stewardship is not ownership**—it is a vocation to protect, nurture, and preserve.

Biblical principles include:

- Sabbath rest for the land (Leviticus 25)

- Reverence for the interconnectedness of all life

- The recognition that creation "groans" under human sin and awaits redemption (Romans 8:22)

Modern Christian movements, especially Catholic (e.g., Laudato Si'), Orthodox, and Evangelical climate initiatives, call for:

- Ecological repentance

- Renewed reverence for God's creation

- Sustainable living as an act of worship

**Environmental Stewardship: Islamic Perspective**

Islam teaches that the earth is a **trust (amānah)** from God, not to be corrupted or misused:

> *"It is He who has appointed you stewards on the earth..."* (Qur'an "Livestock" 6:165)

Nature is filled with signs (āyāt) of God. Animals, rivers, mountains, and skies all exist with purpose and rhythm, reflecting divine design.

Core principles include:

- **Balance** (mīzān): ecological harmony must be preserved

- **Prohibition of waste** (isrāf): overconsumption is spiritually and socially harmful

- Sanctity of life: nature is infused with dignity and **must not be destroyed wantonly**

The Prophet Muhammad emphasized:

- Planting trees as charity

- Protection of water sources

- Kindness to animals and ecosystems

Environmental responsibility is embedded in Islamic rituals—wudu ' (ablution) teaches water conservation; ḥajj includes animal ethics and ecological awareness.

| Criterion | Assessment |
|---|---|
| **1. Communication** | ✔ Connects humanity to creation and one another through shared responsibility. |
| **2. Friction** | ✔ Reduces **resource-based conflict** and geopolitical tension. |
| **3. Problem Solving** | ✔ Promotes **sustainable innovation** and **ethical science**. |
| **4. Resilience** | ✔ Ensures survival through long-term ecological integrity. |
| **5. Trust & Cooperation** | ✔ Builds global solidarity and intergenerational justice. |
| **6. Adaptability** | ✔ Encourages learning from nature and ecological limits. |
| **7. Pro-Social w/o Reward** | ✔ Often involves sacrifice for unseen or future benefits. |

| 8. Functional Health | ✓ Linked to clean air, water, food security, and mental wellness. |

## Environmental Stewardship: Does it Align with Future Human Growth and Evolution?

### Environmental Stewardship: Points of Alignment

- Both Islam and Christianity affirm that creation is sacred, not inert matter.

- Stewardship is seen as a divine trust, not a secular option.

- Care for the environment is not separate from faith—it is a test of sincerity and maturity.

### Environmental Stewardship: Points of Divergence or Nuance

- Islam often frames stewardship within daily practice and legal obligation, encouraging ecological ethics through ritual and law.

- Christianity traditionally emphasized dominion, though modern theology reinterprets this as care rather than control, integrating cosmic redemption into ecological ethics.

- Both faiths are now converging on the view that climate justice is faith in action.

### Environmental Stewardship Conclusion: Evolutionary Trajectory

**Environmental stewardship is not a secondary concern**; it is the evolutionary bedrock upon which all other virtues must stand. Without a habitable planet, there can be no justice, no charity, no worship, only the slow unraveling of everything sacred and civilizational. Both Islam and Christianity remind us that the Earth is not a possession to be exploited, but a trust to be honored, a living testament to divine generosity and a legacy we are bound to protect for those yet unborn. Within the evolutionary-spiritual framework, stewardship of the environment is not simply good ethics; **it is the highest form of moral intelligence**. It reflects a civilization's capacity to align power with reverence, consumption with restraint, and growth with responsibility. In the future shaped by moral evolution, to be truly faithful will mean to be fiercely protective of the soil, the sky, and all that lives between, **not out of fear, but out of love for creation and for the generations whose breath depends on our choices today.**

# Chapter 29: Good Leadership

**Good Leadership: Definition and Purpose**

Good leadership is the exercise of visionary, ethical, and accountable authority in service of others. It is marked by wisdom, humility, justice, and courage, not personal gain or domination. True leadership is not defined by control, but by **service, moral clarity, and the ability to unite people toward the common good.**

In the evolutionary-spiritual framework, good leadership is a central mechanism for moral evolution and societal resilience. It protects the vulnerable, navigates crisis, fosters innovation, and ensures that power is a tool for justice, not tyranny.

**Good Leadership: Christian Perspective**

Jesus redefines leadership entirely:

> *"The greatest among you will be your servant."* (Matthew 23:11)

He models servant leadership—washing feet, forgiving enemies, and dying for others. Leadership in Christianity is:

- Sacrificial: putting others 'needs first

- Prophetic: confronting injustice, not appeasing power

- Pastoral: guiding, nurturing, and protecting the flock

Paul urges leaders to be:

> *"Above reproach... not overbearing, not quick-tempered... but hospitable, self-controlled, upright... "* (Titus 1:6–8)

Christian leaders are stewards of:

- Truth

- Community well-being

- Moral courage in public and private spheres

**Good Leadership: Islamic Perspective**

Leadership (*imāmah, amānah, wilāyah*) in Islam is a sacred trust, not a privilege:

> *"Verily, God commands you to render trusts to whom they are due and when you judge between people to judge with justice."* (Qur'an "The Women" 4:58)

The Prophet Muhammad warned against:

- Leaders who hoard wealth

- Nepotism, oppression, and abuse of power

- Using leadership as a means to self-aggrandizement

He said:

> *"Each of you is a shepherd, and each of you is responsible for his flock."*
> (Bukhari & Muslim)

Ideal Islamic leadership is:

- Just and consultative (*shūrā*)

- Guided by knowledge, humility, and service

- Accountable both to the people and to God

The *Rightly Guided Caliphs (al-Khulafʾā al-Rāshidūn)* are often cited as models of righteous, restrained, and just governance.

| Criterion | Assessment |
|---|---|
| **1. Communication** | ✓ Ethical leadership fosters clarity, empathy, and shared direction. |
| **2. Friction** | ✓ Just leadership reduces corruption, injustice, and unrest. |
| **3. Problem Solving** | ✓ Guides effective, long-term response to complex challenges. |
| **4. Resilience** | ✓ Builds institutional and emotional stability in crisis. |
| **5. Trust & Cooperation** | ✓ Moral leaders inspire loyalty and sustainable teamwork. |
| **6. Adaptability** | ✓ Good leaders **adjust strategy without abandoning values**. |

| | |
|---|---|
| **7. Pro-Social w/o Reward** | ✔ Authentic leaders serve regardless of status or recognition. |
| **8. Functional Health** | ✔ Ethical governance improves public trust and social health. |

## Good Leadership: Does it Align with Future Human Growth and Evolution?

### Good Leadership: Points of Alignment

- Both Islam and Christianity insist that leadership is moral service, not self-service.

- The **leader is not above the law**, but under it, accountable to God, conscience, and community.

- Leadership is a sacred role—demanding virtue, courage, and sacrifice.

### Good Leadership: Points of Divergence or Nuance

- Islam emphasizes legal responsibility and social equity, with a focus on consultation and justice in public governance.

- Christianity emphasizes servant-hearted leadership and are held to a greater accountability.

- Islamic political theory includes juridical guidelines; Christian leadership often operates within spiritual or ecclesial frameworks, especially in democratic or pluralistic states.

### Good Leadership Conclusion: Evolutionary Trajectory

Leadership is not merely a structural necessity; it is a catalyst of moral evolution and a shaping force of collective destiny. Both Islam and Christianity warn that bad leadership does more than mismanage policy; it corrodes public trust, fractures social cohesion, and poisons the ethical soil from which communities grow. Good leadership, by contrast, is among **the most potent evolutionary instruments available to humanity**. It amplifies virtue, corrects systemic injustice, and models a way of being that guides others not by coercion, but by conscience. In the evolutionary-spiritual framework, leadership is no longer about dominance or charisma; it is about ethical transmission. The leader becomes a steward of collective meaning, a guardian of dignity, and a cultivator of justice who serves the sacred within the social. In the future shaped by conscious evolution, the best leaders will not command from above; they will integrate from within, **anchoring their authority in humility, moral clarity, and service to the greater good.**

# Chapter 30: Moral Responsibility and Institutional Trust

**Moral Responsibility and Trust: Definition and Purpose**

Moral responsibility refers to the ethical duty of individuals to act with integrity, accountability, and conscience, especially in relation to social systems and public roles. Institutional trust is the confidence people place in the reliability, fairness, and moral legitimacy of authority structures from family and religious bodies to legal, educational, and governmental systems.

In the evolutionary-spiritual framework, trust and responsibility form the backbone of ethical civilization. They ensure that power serves justice, that rules serve people, and that institutions inspire loyalty not through fear, but through moral credibility.

**Moral Responsibility and Trust: Christian Perspective**

Christian ethics honor both submission to legitimate authority and resistance to unjust power:

*"Let everyone be subject to the governing authorities..."* (Romans 13:1)

*"We must obey God rather than men."* (Acts 5:29)

Jesus models this balance by:

- **Paying taxes and honoring civic duty**

- Rebuking religious leaders and **confronting unjust structures**

- Teaching conscience, service, and **truth over blind conformity**

Paul, Peter, and early Christians endured persecution not through rebellion, but through moral witness, civil obedience where possible, and prophetic resistance when necessary.

Christian traditions today emphasize:

- Discernment in civic responsibility

- Engaged citizenship with spiritual accountability

- Institutional reform rooted in Gospel values

**Moral Responsibility and Trust: Islamic Perspective**

Islam commands obedience to just authority, rooted in service and moral accountability:

> *"O you who believe, obey God and obey the Messenger and those in authority among you…"* (Qur'an "The Women" 4:59)

But this obedience is:

- Conditional on justice: *"There is no obedience in sin."* (Muslim)

- Linked to shūrā (consultation) and amānah (trust):

  > *"Verily, God commands you to render trusts to whom they are due…"* (Qur'an 4:58)

Key concepts:

- **Leaders are servants**, not masters

- Institutions must reflect divine justice and public welfare

- Individuals have a **duty to speak truth to power**, following the Prophetic example

Corruption, abuse of power, and blind obedience are condemned. Moral responsibility includes:

- **Conscientious dissent**

- Community **accountability**

- Balancing submission to order with devotion to truth

| Criterion | Assessment |
|-----------|------------|
| 1. Communication | ✓ Enhances transparency, accountability, and civic dialogue. |
| 2. Friction | ✓ Reduces chaos by **aligning duty with justice**. |
| 3. Problem Solving | ✓ Empowers ethical decision-making in complex systems. |
| 4. Resilience | ✓ Institutions rooted in integrity can withstand crises. |
| 5. Trust & Cooperation | ✓ Ethical leadership fosters loyalty and public cooperation. |
| 6. Adaptability | ✓ Encourages reform without destabilization. |

| | |
|---|---|
| **7. Pro-Social w/o Reward** | ✔ Responsible citizenship often lacks applause but sustains the moral order. |
| **8. Functional Health** | ✔ Reduces corruption, inefficiency, and cynicism. |

**Responsibility and Trust: Does it Align with Future Human Growth and Evolution?**

**Moral Responsibility and Trust: Points of Alignment**

- Both Islam and Christianity affirm that **true authority must serve truth**, justice, and the common good.

- **Obedience is not absolute**, but rooted in discernment, righteousness, and sacred trust.

- Individuals must practice moral responsibility as both citizens and believers.

**Moral Responsibility and Trust: Points of Divergence or Nuance**

- Islam tends to institutionalize ethical governance, providing structured guidance through Sharīʿah and prophetic tradition.

- Christianity offers more theological frameworks for conscience-driven resistance, especially under unjust regimes.

- Both traditions converge on the principle that legitimate authority must reflect divine justice, and that blind obedience is morally dangerous.

**Moral Responsibility and Trust Conclusion: Evolutionary Trajectory**

Authority without ethics breeds oppression, conscience without connection breeds fragmentation. Islam and Christianity both insist that **trust must be earned**, and that **obedience must be discerning**, anchored not in fear, but in justice. In the evolutionary-spiritual framework, moral responsibility and institutional trust must rise together as twin pillars of an ethically advanced society. The most resilient civilizations will be those where institutions are transparent and principled, and where individuals do not retreat into cynicism or revolt into chaos but participate in truth-seeking with integrity and discernment. In such a future, leadership will no longer rest on power alone, and law will no longer suffice without legitimacy. **Morality will not be enforced from above as mere compliance; it will be embodied from within as collective wisdom**. This is the evolutionary shift: from systems that manage behavior to cultures that cultivate character, where the social contract is not imposed but internalized as sacred trust.

# Chapter 31: Interfaith Respect

**Interfaith Respect: Definition and Purpose**

Interfaith respect is the moral commitment to recognize the dignity, sincerity, and spiritual value of people from different religious traditions, **even when one does not agree with their theology**. It promotes peace, dialogue, and shared moral action, especially in pluralistic societies where diversity is a daily reality.

In the evolutionary-spiritual framework, interfaith respect is a hallmark of advanced moral maturity. It allows for cooperation across difference, guards against sectarian violence, and fosters a sacred pluralism that values truth-seeking over tribal supremacy.

**Interfaith Respect: Christian Perspective**

Jesus exemplifies compassionate engagement across religious and ethnic boundaries:

- Speaking with the Samaritan woman (John 4)

- Healing a Roman centurion's servant (Matthew 8)

- Praising the "Good Samaritan" (Luke 10)

Paul teaches:

> *"If possible, as far as it depends on you, live at peace with everyone."* (Romans 12:18)

Christian theology affirms:

- Love of neighbor is unconditional

- Truth and grace must be offered without arrogance

- Dialogue is a form of witness and humility, not conquest

The Second Vatican Council (*Nostra Aetate*) marked a major turn in Catholic doctrine, affirming:

- Reverence for truth and holiness found in other religions

- Condemnation of religious hatred

- Commitment to interfaith friendship and cooperation

**Interfaith Respect: Islamic Perspective**

The Qur'an affirms that **religious diversity is part of divine will**:

*"To each of you We have prescribed a law and a way. Had God willed, He could have made you one nation..."* (Qur'an "The Table Spread" 5:48)

Key teachings:

- **No compulsion in religion** (Qur'an "The Cow" 2:256)

- Respectful dialogue with People of the Book (Qur'an "The Spider" 29:46)

- Justice and good conduct even toward non-Muslims (Qur'an "She that is to be Examined" 60:8)

The Prophet Muhammad:

- Established treaties with Jewish and Christian communities

- Prohibited insults against other religions

- Promoted mutual dignity and civic cooperation

While Islam maintains theological distinctions, it calls for peaceful coexistence, ethical partnership, and **shared concern for justice**.

| Criterion | Assessment |
|---|---|
| **1. Communication** | ✔ Enables dialogue, mutual learning, and respectful coexistence. |
| **2. Friction** | ✔ Reduces bigotry, suspicion, and intergroup violence. |
| **3. Problem Solving** | ✔ Allows for collaboration on shared moral and global challenges. |
| **4. Resilience** | ✔ Pluralistic societies with respect are more stable and peaceful. |
| **5. Trust & Cooperation** | ✔ Builds bridges across theological and cultural divides. |
| **6. Adaptability** | ✔ Encourages openness and mutual evolution in understanding. |

| | |
|---|---|
| **7. Pro-Social w/o Reward** | ✔ Respect is extended even without reciprocity or agreement. |
| **8. Functional Health** | ✔ Correlates with lower extremism and better civic cohesion. |

## Interfaith Respect: Does it Align with Future Human Growth and Evolution?

### Interfaith Respect: Points of Alignment

- Both Islam and Christianity teach that respect for the religious "other" is a measure of faithfulness to God's mercy and justice.

- Interfaith respect is not relativism—it is humility before the mystery of divine wisdom, and a refusal to reduce people to their categories.

- The moral imperative is to seek peace, share moral vision, and collaborate on justice and mercy.

### Interfaith Respect: Points of Divergence or Nuance

- Islam asserts the finality of the Qur'an, while still affirming shared truths in previous revelations.

- Christianity proclaims salvation through Christ yet increasingly acknowledges the presence of grace and truth in other traditions.

- While theological boundaries remain, both traditions are evolving toward dialogue-based missions of love, dignity, and mutual service.

### Interfaith Respect Conclusion: Evolutionary Trajectory

Interfaith respect is not just the etiquette of civility; **it is the evolutionary scaffolding of a pluralistic, morally resilient world**. Without it, difference hardens into division and belief becomes a battleground; with it, diversity becomes dialogue, and conviction becomes a catalyst for connection. Both Islam and Christianity affirm that truth does not weaken in the face of openness, and that reverence for another's path need not compromise one's own. In the evolutionary-spiritual framework, interfaith respect is a sign of **ethical maturity**, a recognition **that spiritual wisdom is not monopolized**, but manifold. Those who build bridges across belief are not diluting truth, but refining it, choosing peace over pride and unity over uniformity. In the future shaped by moral evolution, sacred pluralism will not be seen as a threat to fidelity, but as its deepest expression: **the ability to love without borders, to uphold justice without prejudice, and to walk humbly alongside those whose path may differ but whose purpose converges in the pursuit of the Divine.**

# Chapter 32: Grave Sins and Accountability

### Grave Sins and Accountability: Definition and Purpose

Grave sins are serious moral violations that cause deep harm to self, others, or the social and spiritual order. These are not trivial mistakes but acts of willful injustice, cruelty, deceit, or neglect. Accountability is the process of recognizing, confronting, and taking responsibility for these wrongs, whether through repentance, reparation, or justice.

In the evolutionary-spiritual framework, confronting grave sin is essential for moral maturity and communal resilience. A society that ignores deep wrongs cannot sustain trust or dignity. **A soul that avoids accountability cannot grow**. Redemption begins where denial ends.

### Grave Sins and Accountability: Christian Perspective

Christianity also identifies grave sins as willful breaches of God's law and love. Traditional categories include:

- Mortal sins (Catholic tradition): destroy grace and require confession and repentance

- Sins of omission: failure to act when justice demands it

Paul teaches:

*"The wages of sin is death, but the gift of God is eternal life..."* (Romans 6:23)

Jesus calls for:

- Heart-level repentance (Matthew 5–7)

- Forgiveness of others as a condition of divine forgiveness

- Moral courage in naming and confronting sin (Matthew 18)

Christian theology insists that:

- **No sin is beyond redemption**, but

- Grace requires confession, humility, and change

**Grave Sins and Accountability: Islamic Perspective**

Islam classifies major sins (*kab'āir*) as those that result in severe consequences in this life or the next. These include:

- Murder, theft, adultery, false testimony

- Oppression, neglecting orphans, pride, usury

- Breaking oaths, dishonoring parents, hypocrisy

   *"If you avoid the major sins which you are forbidden, We will remove from you your lesser sins..."* (Qur'an "The Women" 4:31)

Yet Islam offers:

- Mercy through repentance (tawbah):

   *"Do not despair of the mercy of God..."* (Qur'an 39:53)

- A divine balance of justice and forgiveness

- Emphasis on self-awareness, moral vigilance, and legal accountability

The Prophet Muhammad said:

*"All of the children of Adam are sinners, and the best of sinners are those who repent."* (Tirmidhi)

| Criterion | Assessment |
|---|---|
| **1. Communication** | ✔ Truthful confession restores integrity and mutual respect. |
| **2. Friction** | ✔ Accountability prevents cycles of blame and retaliation. |
| **3. Problem Solving** | ✔ Naming wrongdoing enables healing, justice, and reform. |
| **4. Resilience** | ✔ Societies that face failure honestly emerge stronger. |
| **5. Trust & Cooperation** | ✔ Ethical systems demand trust grounded in moral responsibility. |
| **6. Adaptability** | ✔ Accountability allows for change, evolution, and healing. |
| **7. Pro-Social w/o Reward** | ✔ Owning faults often involves shame or sacrifice but yields long-term virtue. |

| | |
|---|---|
| **8. Functional Health** | ✓ Reduces hidden guilt, moral anxiety, and institutional collapse. |

## Grave Sins and Accountability: Does it Align with Future Human Growth and Evolution?

### Grave Sins and Accountability: Points of Alignment

- Both Islam and Christianity insist that grave wrongs must be confronted, not concealed.

- Mercy and accountability are not opposites—they are two wings of moral flight.

- Divine judgment is always tempered with divine mercy, but truth precedes healing.

### Grave Sins and Accountability: Points of Divergence or Nuance

- Islam links accountability to public ethics and law, while Christianity emphasizes personal confession and spiritual integrity.

- Christianity roots transformation in grace, while Islam insists on intentional repentance, restitution, and reform.

- Both traditions affirm that the **weight of sin can be lifted but not ignored**.

### Grave Sins and Accountability Conclusion: Evolutionary Trajectory

A healthy soul and a healthy civilization must learn to confront harm with honesty and humility. Grave sins, whether personal betrayals or systemic injustices, do not fade with time; when left unaddressed, they become graves for compassion, conscience, and communal trust. Islam and Christianity both call us not to perfection, but to responsibility, to face our failures not with despair, but with the courage to repent, restore, and repair. In the evolutionary-spiritual framework, accountability is not a punishment; it is a portal. It marks the shift from primitive denial and moral stagnation to higher forms of ethical resilience and collective healing. Those who practice truth-telling and responsibility are not weakening society; they are safeguarding its future, ensuring that wounds are not buried but transformed. **In the moral evolution of humanity, light does not come by avoiding darkness, but by walking through it with clarity, humility, and the will to make what is broken whole again.**

# Sacred Evolution

# Part IV: Spiritual Practices and Principles

# Chapter 33: Hope and Endurance

### Hope and Endurance: Definition and Purpose

Hope is the confident trust in the possibility of goodness, meaning, or redemption, even when unseen. Endurance is the sustained strength to remain faithful, moral, and resilient despite hardship, delay, or uncertainty. Together, they equip the soul and society to withstand suffering without moral collapse, and to act toward the good even when the outcome is uncertain.

In the evolutionary-spiritual model, hope and endurance are essential for long-term flourishing. They prevent despair from becoming nihilism and ensure that ethical action can survive trial and time. Civilizations don't just need cleverness—they need courage grounded in purpose.

### Hope and Endurance: Christian Perspective

Hope is one of the three theological virtues (1 Corinthians 13:13), often paired with faith and love. It is rooted in:

- The resurrection of Christ

- The promise of redemption and ultimate justice

- The ongoing presence of the Spirit in suffering

Paul writes:

*"We also glory in our sufferings, because we know that suffering produces perseverance; perseverance, character; and character, hope."*
(Romans 5:3–4)

Christian hope includes:

- Eschatological vision: the kingdom to come

- Daily endurance: trusting in God's goodness now

- Cross-centered strength: modeling endurance after Christ's passion

Hope is not optimism—it is the defiant trust that God's goodness will prevail, even when hidden.

## Hope and Endurance: Islamic Perspective

Hope (*raj'a*) is central in Islam, always paired with tawakkul (trust in God) and ṣabr (patience).

> *"Do not despair of the mercy of God. Indeed, God forgives all sins."* (Qur'an "The Crowds" 39:53)

The believer is called to:

- **Hope in divine mercy**, even after sin or failure

- Endure trials without resentment

- Maintain moral and spiritual effort, **even when justice is delayed**

The Prophet Muhammad's endurance:

- Persecuted in Mecca, boycotted, exiled, yet unwavering

- Built community in Medina while facing war and internal strife

- Never allowed hardship to displace hope or corrupt integrity

Hope in Islam is not passive—it is an act of moral vision rooted in divine promise.

| Criterion | Assessment |
|---|---|
| **1. Communication** | ✓ Enables resilient dialogue in times of hardship and disagreement. |
| **2. Friction** | ✓ Helps process pain without blame, and conflict without despair. |
| **3. Problem Solving** | ✓ Sustains moral clarity and perseverance across slow, complex challenges. |
| **4. Resilience** | ✓ Anchors emotional and communal strength through adversity. |
| **5. Trust & Cooperation** | ✓ Hope reinforces loyalty and shared moral purpose. |
| **6. Adaptability** | ✓ Offers inner stability while allowing flexible response to change. |
| **7. Pro-Social w/o Reward** | ✓ Hopeful endurance often persists with no visible success. |

| 8. Functional Health | ✔ Strong predictor of emotional well-being and mental health. |
|---|---|

## Hope and Endurance: Does it Align with Future Human Growth and Evolution?

### Hope and Endurance: Points of Alignment

- Both Islam and Christianity teach that hope is essential for endurance, and that endurance perfects the soul.

- Hope is not naive belief in comfort, but mature trust in the enduring worth of goodness.

- Endurance is framed as moral stamina, not just stoic survival.

### Hope and Endurance: Points of Divergence or Nuance

- Islam links hope closely to mercy, patience, and divine justice, with practical rituals for renewing hope (e.g., prayer, dhikr, du'ā').

- Christianity often links hope to Christ's resurrection and eternal life, and views endurance through the lens of spiritual transformation.

- Islam emphasizes trust in God's decree with effort, while Christianity emphasizes grace as sustaining power through weakness.

### Hope and Endurance Conclusion: Evolutionary Trajectory

Hope and endurance are the twin engines of ethical evolution, the forces that keep the human spirit intact when certainty dissolves and suffering deepens. They empower us to persevere without bitterness and to act with integrity even when justice is delayed or unseen. Both Islam and Christianity call us to **resist despair**, to **rise above spiritual fatigue**, and to trust that **struggle is not the enemy of faith**, but its proving ground. In the evolutionary-spiritual framework, hope is not escapism; it is a strategic virtue that sustains resilience over time and guides moral vision beyond the visible horizon. Endurance, likewise, becomes more than survival; it is the discipline of continuing to live with purpose, even when meaning feels out of reach. In the future shaped by conscious evolution, it will be those who endure with hope, rather than cynicism, who will carry the sacred forward, proving that meaning is not erased by pain, but revealed through it, refined within it, and renewed because of it.

# Chapter 34: The Ten Commandments

## The Ten Commandments: Definition and Purpose

The Ten Commandments are a set of foundational moral directives revealed to Moses on Mount Sinai, forming a core ethical structure for Judaism, Christianity, and, by extension, Islamic ethics. They offer a succinct framework for right relationship with God and with others, emphasizing moral law, sacred obligation, and social harmony.

In the evolutionary-spiritual framework, the Ten Commandments represent a primal ethical code that undergirds cooperative survival. They encode principles of trust, restraint, justice, and fidelity, which are essential for maintaining interpersonal integrity and communal cohesion.

## The Ten Commandments: Christian Perspective

The Ten Commandments are found in Exodus 20 and Deuteronomy 5, divided into:

- Commands 1–4: governing relationship with God

- Commands 5–10: governing relationship with others

Jesus reinterprets and intensifies them in the Sermon on the Mount, summarizing them under two great commandments:

"Love the Lord your God with all your heart..."

"Love your neighbor as yourself." (Matthew 22:37–39)

Key principles include:

- Sacred reverence (no idols, honoring the name of God)

- Sabbath rest

- Respect, nonviolence, marital fidelity, honesty, and contentment

Christian ethics emphasize internalization of the law:

- From external compliance to inner transformation

- From legalistic observance to love-based fulfillment

**The Ten Commandments: Islamic Perspective**

While Islam does not present the Ten Commandments as a single list, all their moral teachings are affirmed across the Qur'an and Hadith.

For example:

- Monotheism: *"There is no god but God."* (Qur'an "Muhammad" 47:19)

- Respect for parents: *"And be kind to parents."* (Qur'an "The Night Journey" 17:23)

- Prohibition of murder: *"Whoever kills a soul... it is as if he had slain mankind entirely."* (Qur'an "The Table Spread" 5:32)

- Condemnation of theft, adultery, false witness, and envy: addressed extensively across Qur'anic legal and spiritual teachings

Islamic ethics integrate these values into:

- Legal systems (ḥudūd laws)

- Daily spiritual life (sincerity in worship, justice in dealings)

- Societal norms rooted in accountability to both God and community

The essence of the Ten Commandments is preserved in Islam, expressed through the unity of law, devotion, and ethical discipline.

| Criterion | Assessment |
|---|---|
| **1. Communication** | ✔ Creates a shared ethical language and moral clarity. |
| **2. Friction** | ✔ Regulates behavior to minimize interpersonal and social harm. |
| **3. Problem Solving** | ✔ Establishes justice, trust, and consistency in group dynamics. |
| **4. Resilience** | ✔ Anchors moral behavior during trial, temptation, or cultural change. |
| **5. Trust & Cooperation** | ✔ Sets a moral baseline for relational reliability. |
| **6. Adaptability** | ✔ Timeless core principles applicable across contexts, if interpreted wisely. |
| **7. Pro-Social w/o Reward** | ✔ Encourages duty and virtue even without surveillance or recognition. |

| 8. Functional Health | ✔ Prevents destructive behaviors, fosters family and civic stability. |
|---|---|

## Ten Commandments: Does it Align with Future Human Growth and Evolution?

### The Ten Commandments: Points of Alignment

- Islam and Christianity both affirm the core moral content of the Ten Commandments.

- Both view these principles as divinely revealed, morally binding, and spiritually elevating.

- The commandments foster social trust, spiritual awareness, and personal accountability.

### The Ten Commandments: Points of Divergence or Nuance

- Christianity often emphasizes the symbolic and covenantal nature of the Ten Commandments within the narrative of salvation.

- Islam integrates these principles into a broader legal and ethical framework, focusing on justice, worship, and community rights.

- Sabbath observance is not central in Islam, but spiritual rhythm is preserved through daily prayer, weekly Jumu'ah, and annual fasting.

### The Ten Commandments Conclusion: Evolutionary Trajectory

The Ten Commandments are not fossilized relics of a bygone era; they are enduring signposts in humanity's ethical evolution. They represent a moral architecture that has allowed communities to stabilize, relationships to thrive, and the sacred to find expression in everyday life. Islam and Christianity both affirm these principles as foundational to divine justice and human dignity. In the evolutionary-spiritual framework, the value of the Ten Commandments lies not in rigid legalism, but in their **capacity to align the human conscience** with the deeper rhythms of covenant, restraint, and reverence. These laws are **not arbitrary**; they are adaptive wisdom, shaping the conditions for collective survival, moral responsibility, and spiritual coherence. In the future shaped by ethical evolution, the spirit of the Commandments will continue **to guide humanity, not by coercion, but by inner conviction**, curbing the ego's excesses and honoring the sacred within both human society and the cosmos we inhabit.

# Chapter 35: Sabbath and Rest

**Sabbath and Rest: Definition and Purpose**

The Sabbath is a sacred rhythm of ceasing from labor, traditionally observed once a week as a time for spiritual renewal, reflection, and relational restoration. More broadly, rest is the theological and ethical acknowledgment that human life requires balance, limits, and grace, a deliberate interruption in productivity to reconnect with purpose.

In the evolutionary-spiritual framework, rest is not indulgence; **it is a biological, emotional, and moral necessity**. Societies and individuals that rest wisely are more resilient, focused, creative, and compassionate. Sabbath is how the soul breathes.

**Sabbath and Rest: Christian Perspective**

The Sabbath originates in Genesis, where God rests on the seventh day (Genesis 2:2–3). The Ten Commandments institutionalize it:

*"Remember the Sabbath day, to keep it holy. Six days you shall labor..."*
(Exodus 20:8–10)

Jesus reinterprets the Sabbath not as legalism but liberation:

*"The Sabbath was made for man, not man for the Sabbath."* (Mark 2:27)

Sabbath in Christian tradition includes:

- Rest from labor

- Worship and gratitude

- Family and community reconnection

- Spiritual rejuvenation through grace

While Sunday worship replaced the Saturday Sabbath in most traditions, the principle of rest remains central. Monastic traditions elevate rest as sacred rhythm, while modern movements call for Sabbath as resistance— a protest against materialism, burnout, and performance culture.

**Sabbath and Rest: Islamic Perspective**

Islam does not observe a Sabbath in the legalistic sense found in Judaism, but it does sanctify rest and sacred time, particularly on Fridays (Jumu'ah):

> *"O you who believe! When the call is made for prayer on Friday, hasten to the remembrance of God and leave off trade..."* (Qur'an 62:9)

Islamic rest is:

- Spiritual: through prayer (*ṣalāh*), recitation, and remembrance (*dhikr*)

- Rhythmic: through **daily cycles of prayer** and annual fasts (e.g., Ramadan)

- Balanced: work and rest are both acts of worship when properly aligned

The Prophet Muhammad emphasized:

- Mercy toward oneself and others

- Rest between acts of devotion

- Integration of work with reverent stillness, not frenetic excess

Islamic theology teaches that overwork and wastefulness are both spiritual imbalances, and that moderate living leads to clarity and nearness to God.

| Criterion | Assessment |
|---|---|
| **1. Communication** | ✔ Deepens connection with self, others, and the sacred. |
| **2. Friction** | ✔ Reduces stress, burnout, and relational neglect. |
| **3. Problem Solving** | ✔ Enables creative clarity and moral reflection. |
| **4. Resilience** | ✔ Provides recovery cycles for long-term stamina. |
| **5. Trust & Cooperation** | ✔ Reorients social rhythms toward shared sacred values. |
| **6. Adaptability** | ✔ Culturally flexible while maintaining spiritual principle. |
| **7. Pro-Social w/o Reward** | ✔ Rest is non-productive by design, but deeply formative. |

| | |
|---|---|
| **8. Functional Health** | ✓ Correlated with mental wellness, longevity, and family strength. |

**Sabbath and Rest: Does it Align with Future Human Growth and Evolution?**

**Sabbath and Rest: Points of Alignment**

- Both Islam and Christianity affirm that sacred rest is essential to moral and spiritual health.

- Rest is not just recuperation—it is a ritual of alignment with the sacred order.

- Proper rest restores the soul, reorients desire, and recommits the heart to what truly matters.

**Sabbath and Rest: Points of Divergence or Nuanc**

- Islam does not designate a full day of rest but builds rest into daily rituals, especially the Jumuʿah prayer, which re-centers the week.

- Christianity sets aside the Sabbath as a theological institution, emphasizing God's rest and human liberation.

- Both offer rest as a response to divine love—not escapism, but reverent recharging.

**Sabbath and Rest Conclusion: Evolutionary Trajectory**

A world that never stops will inevitably forget what it means to be human. In both Islam and Christianity, we are called to remember the sacredness of stillness, to resist the relentless drive toward productivity, and to honor our bodies, minds, and souls through rhythms of rest. In the evolutionary-spiritual framework, the Sabbath is not merely a divine suggestion; it is a civilizational necessity. It **reintroduces balance** into a world bent on acceleration, reminding us that growth without pause leads to burnout, and progress without reflection leads to collapse. The practice of rest is not withdrawal from life; it is how life becomes meaningful again. It signals to ourselves and our communities that we are not machines to be optimized, but moral beings shaped by silence, presence, and reverence. In the future shaped by ethical evolution, Sabbath will be reclaimed not only as ritual, but as resistance, **an act of spiritual defiance against the cult of endless doing, and a sacred return to being.**

# Chapter 36: Eschatology and the Moral Arc of History

**Eschatology: Definition and Purpose**

Eschatology is the theological study of final things—death, judgment, the afterlife, and the ultimate destiny of the world. It offers a cosmic vision of justice, affirming that history has direction, consequence, and sacred culmination. It links present ethics to future accountability, and suffering to eventual redemption.

In the evolutionary-spiritual framework, eschatology supplies a deep moral horizon. It grounds conscience in cosmic scale, encourages long-term responsibility, and ensures that truth, love, and justice are not fleeting ideals, but eternal trajectories.

**Eschatology: Christian Perspective**

Christian eschatology centers on the Second Coming of Christ, final judgment, resurrection, and the restoration of all creation:

> *"He will come again in glory to judge the living and the dead."* (Nicene Creed)

Themes include:

- Heaven and hell as eternal realities

- The New Jerusalem: a redeemed world of peace and righteousness (Revelation 21)

- Divine justice fulfilled and evil ultimately defeated

Jesus teaches readiness:

*"Keep watch, for you do not know the day or the hour."* (Matthew 25:13)

Paul emphasizes:

- Moral preparation through love and purity

- Hope of resurrection through Christ

- A vision of history shaped by grace, not fatalism

Eschatology is both personal (death, judgment) and global (the moral arc of humanity).

## Eschatology: Islamic Perspective

Belief in the Last Day (Yawm al-Qiyāmah) is one of Islam's six articles of faith. The Qur'an is filled with descriptions of:

- Resurrection of the dead

- Divine judgment

- Paradise (Jannah) and Hell (Jahannam)

- The scale (mīzān) and the record of deeds (kitāb)

*"Every soul shall taste death. Then to Us you will be returned."* (Qur'an "The Spider" 29:57)

Key theological themes include:

- Moral accountability: no deed is forgotten

- Mercy and justice in balance

- **Cosmic renewal, not annihilation**—heaven and earth will be reformed

Prophetic traditions detail:

- Signs of the Last Hour

- The return of Jesus ('Isa), the rise of the Mahdī, and the defeat of falsehood (Dajjāl)

Eschatology in Islam is not speculative—it is a **motivation for action**, a call to ethical urgency, and a reminder that time is sacred.

| Criterion | Assessment |
|---|---|
| **1. Communication** | ✔ Gives language to final purpose and moral consequence. |
| **2. Friction** | ✔ **Reduces despair** and revenge by framing justice cosmically. |
| **3. Problem Solving** | ✔ Sustains moral effort when earthly outcomes fall short. |
| **4. Resilience** | ✔ Offers **hope** in persecution, loss, or apparent failure. |
| **5. Trust & Cooperation** | ✔ Encourages faith in ultimate justice over vigilantism. |

| | |
|---|---|
| **6. Adaptability** | ⚠ Literalism can hinder flexibility; symbolic vision enhances growth. |
| **7. Pro-Social w/o Reward** | ✔ Inspires ethical action with or without temporal success. |
| **8. Functional Health** | ✔ Reduces existential anxiety and offers moral orientation. |

## Eschatology: Does it Align with Future Human Growth and Evolution?

### Eschatology: Points of Alignment

- Both Islam and Christianity teach that history is not meaningless—it is moral and sacred.

- Eschatology encourages ethical vigilance, humility, and hope beyond circumstance.

- Divine justice is not yet finished, but faithfulness now shapes what is to come.

### Eschatology: Points of Divergence or Nuance

- Islamic eschatology is detail-rich, with vivid imagery and structured sequence; Christianity includes both symbolic and literal strands.

- Islam emphasizes judgment through balance (mīzān); Christianity emphasizes grace through Christ.

- Both traditions emphasize repentance, preparation, and the eternal stakes of present choices.

**Eschatology Conclusion: Evolutionary Trajectory**

Eschatology reminds humanity that we are more than dust; we are destined, not merely to end, but to fulfill. It affirms that morality matters, that history has direction, and that love, justice, and truth are not fleeting ideals but eternal anchors. Both Islam and Christianity proclaim that the end is not annihilation; it is awakening, a moment of reckoning, mercy, and renewal. In the evolutionary-spiritual framework, **eschatology is no longer feared as an apocalyptic rupture**, but revered as a moral horizon, a sacred arc that bends toward justice, compelling us to live as those who will be remembered, judged, and ultimately restored. It reframes the future not as a void, but as the culmination of ethical evolution, where every act of compassion and every choice of integrity becomes part of a deeper unfolding. In this future, eschatology will guide not our speculation, but our transformation, **calling us to live with urgency, humility, and cosmic responsibility**, as participants in a story that transcends us yet depends on us.

# Chapter 37: Freedom of Conscience

**Freedom of Conscience: Definition and Purpose**

Freedom of conscience is the right and moral capacity of each individual to think, believe, and act **in accordance with their inner convictions—** especially in matters of faith, ethics, and meaning. It is not a license to do whatever one pleases, but the **protected space to follow one's moral compass without coercion.**

In the evolutionary-spiritual framework, freedom of conscience is essential for moral creativity, social peace, and spiritual authenticity. It protects the individual from tyranny, the society from uniformity, and the sacred from compulsion. **It is the foundation of a just and pluralistic civilization.**

**Freedom of Conscience: Christian Perspective**

Jesus honored individual freedom repeatedly:

- Inviting, not compelling: *"If anyone would come after me..."* (Matthew 16:24)

- Emphasizing inner transformation, not outer compliance

- Teaching that faith must be freely chosen, as in the parable of the rich young ruler who walks away

Paul affirms the primacy of conscience:

*"Each of them should be fully convinced in their own mind."* (Romans 14:5)

Throughout history, Christianity has oscillated between defending conscience (e.g., early martyrs) and violating it (e.g., forced conversions, inquisitions). Today, most Christian traditions affirm:

- Religious liberty as a divine gift

- **Freedom of conscience as essential to faith**

- Dialogue, not domination, as the method of witness

**Freedom of Conscience: Islamic Perspective**

The Qur'an declares unequivocally:

*"There is no compulsion in religion. Truth has become clear from error."*
(Qur'an "The Cow" 2:256)

Islam holds that true **faith cannot be forced**. It must emerge from sincerity and clarity. Additional teachings include:

- The primacy of intention (niyyah) in all acts of faith and morality

- The **right of "People of the Book"** (Jews and Christians) to religious autonomy

- Historical precedent: the Constitution of Medina, which guaranteed religious freedom and civic cooperation

Classical Islamic jurisprudence did debate limitations around apostasy and heresy, often **distinguishing between private belief and public rebellion**. However, contemporary scholars increasingly affirm:

- The **right to question**, wrestle, and seek

- The principle that **God alone judges** the heart

- The call to invite with wisdom, **not force** (Qur'an 16:125)

| Criterion | Assessment |
|---|---|
| **1. Communication** | ✔ Enables authentic dialogue and moral discernment. |
| **2. Friction** | ✔ Reduces religious violence and ideological coercion. |
| **3. Problem Solving** | ✔ Encourages pluralism, humility, and creative synthesis. |
| **4. Resilience** | ✔ Protects conscience under pressure and social change. |
| **5. Trust & Cooperation** | ✔ Builds mutual respect across differences. |
| **6. Adaptability** | ✔ Vital for coexistence in diverse, modern societies. |
| **7. Pro-Social w/o Reward** | ✔ Conscience often leads one to act rightly without approval. |
| **8. Functional Health** | ✔ Linked to mental peace, reduced dogmatism, and ethical clarity. |

**Freedom of Conscience: Does it Align with Future Human Growth and Evolution?**

**Freedom of Conscience: Points of Alignment**

- Both Islam and Christianity affirm that **faith must be freely chosen to be authentic**.

- Coercion is spiritually counterproductive—only sincerity leads to salvation.

- **Conscience is seen as a gift from God** and a mark of moral dignity.

**Freedom of Conscience: Points of Divergence or Nuance**

- Islam historically tied conscience to public order, with debates around apostasy and communal cohesion, though modern Islamic thought increasingly supports broad conscience rights.

- Christianity evolved from being a persecuted minority (emphasizing conscience) to imperial religion (often restricting it)—but is now reclaiming its roots in moral freedom.

- Both traditions recognize the tension between authority and conscience, and the need for ethical courage to resolve it.

**Freedom of Conscience Conclusion: Evolutionary Trajectory**

Freedom of conscience is not just a political right; it is a cornerstone of spiritual evolution and ethical maturity. It safeguards the soul's journey toward truth and preserves society's capacity to grow without coercion or dogmatic stagnation. Both Islam and Christianity affirm that God desires willing hearts, not ideological hostages, and that **authentic faith cannot be forced**, and **moral insight cannot be inherited without question**. In the evolutionary-spiritual framework, conscience is understood as an inner faculty of discernment, a sacred space where truth is not imposed from above, but discovered through **reflection, humility, and lived** experience. In the future shaped by ethical evolution, freedom of conscience will not merely be tolerated; it will be cultivated as one of humanity's most sacred trusts. For only where conscience is free can moral truth evolve, where hearts are open can love be genuine, and **where thought is unshackled can faith become transformative rather than transactional**.

# Chapter 38: Repentance

## Repentance: Definition and Purpose

Repentance is the spiritual and moral process of acknowledging wrongdoing, feeling sincere remorse, and committing to ethical and behavioral transformation. It is not merely regret—it is the turning of the self toward healing, both inwardly and in relation to others and to God.

In the evolutionary-spiritual framework, repentance is the **key to moral resilience**. It allows individuals and communities to correct course without collapse, repair trust, and transform failure into growth. Without repentance, guilt festers or denial spreads; with it, the past is not erased, but redeemed.

## Repentance: Christian Perspective

Repentance (*metanoia*) means a change of mind and heart—a turning away from sin and toward God. It is foundational in Jesus 'teaching:

> *"Repent, for the kingdom of heaven has come near."* (Matthew 4:17)

Christian repentance includes:

- Confession of sin

- Sorrow and contrition

- Faith in God's forgiveness through Christ

- Renewed obedience and ethical living

In Catholic tradition, this is often practiced through the Sacrament of Reconciliation. In Protestant traditions, it is emphasized as:

- Personal sincerity

- Spiritual renewal

- Fruit-bearing transformation

Repentance is not just for conversion. It is a daily rhythm of humility and renewal.

**Repentance: Islamic Perspective**

Repentance (*tawbah*) is a central theme in the Qur'an and Hadith, tied to God's mercy and the human condition:

*"And turn to God in repentance, all of you, O believers, that you might succeed."* (Qur'an "The Light" 24:31)

God is Al-Tawwāb (The Accepter of Repentance), and no sin is beyond His forgiveness if one returns sincerely.

The conditions of true repentance are:

1. Cease the sin

2. Feel sincere remorse

3. Resolve not to return to it

4. Make amends if the harm affected others

The Prophet Muhammad said:

*"The one who repents from sin is like one who never sinned."* (Ibn Mājah)

Repentance in Islam is:

- Direct (no intermediary needed)

- Immediate (encouraged without delay)

- Ongoing (a lifelong practice)

It is seen as a sign of moral courage, not weakness.

| Criterion | Assessment |
|---|---|
| **1. Communication** | ✔ Reopens moral dialogue and restores honesty. |
| **2. Friction** | ✔ Heals interpersonal wounds and prevents cycles of harm. |
| **3. Problem Solving** | ✔ Allows for learning and ethical recalibration. |
| **4. Resilience** | ✔ Builds inner strength by confronting failure and moving forward. |
| **5. Trust & Cooperation** | ✔ Rebuilds broken trust through contrition and change. |
| **6. Adaptability** | ✔ Encourages moral flexibility and spiritual growth. |

| | |
|---|---|
| **7. Pro-Social w/o Reward** | ✔ Requires courage and humility, often without praise. |
| **8. Functional Health** | ✔ Reduces guilt, anxiety, and internal fragmentation. |

## Repentance: Does it Align with Future Human Growth and Evolution?

### Repentance: Points of Alignment

- Both Islam and Christianity teach that repentance is necessary for forgiveness and transformation.

- It is **not shame-based**—it is **hope-based**: the belief that the human heart can change, and that God welcomes the return.

- Repentance is a path of dignity, restoring brokenness into a higher form of wholeness.

### Repentance: Points of Divergence or Nuance

- Islam emphasizes direct repentance to God, structured by intent, accountability, and reparation.

- Christianity emphasizes repentance through grace, often centered on Christ's atonement and the power of spiritual renewal.

- Catholicism ritualizes repentance sacramentally; Protestantism and Islam stress personal, internal sincerity.

### Repentance Conclusion: Evolutionary Trajectory

Repentance is the moral immune system of the evolving soul, a sacred process by which failure is metabolized into formation. **Without it,**

**wounds fester and patterns repeat**; with it, brokenness becomes the birthplace of wisdom. Both Islam and Christianity teach that no fall is final, and no sin greater than divine mercy. What ultimately matters is not how low we have descended, but how deeply we are willing to rise, with humility, resolve, and the courage to be remade. In the evolutionary-spiritual framework, repentance is not guilt management; **it is adaptive growth**. It is the interior discipline that allows individuals and societies to **confront harm without denial**, to **convert error into transformation**, and to realign with truth in the aftermath of rupture. In each of the traditions, true repentance is fundamentally **a recognition of the violation of a principal**, and not just an act committed. Therefore, meaningful repentance is a **commitment to reaffirm the value of a violated principal**, with a commitment to avoid a repetitive failure. In the future shaped by conscious evolution, repentance will be seen not as weakness, but as strength, the sacred capacity to evolve toward the good, **to integrate the past without being imprisoned** by it, and to turn pain into purpose.

# Chapter 39: Covenant and Trust

**Covenant and Trust: Definition and Purpose**

A covenant is a solemn moral and spiritual agreement, often sacred, between individuals, communities, or between humanity and God. Trust is the ethical glue that makes such commitments durable, the confidence that promises will be kept, responsibilities upheld, and intentions honored.

In the evolutionary-spiritual framework, covenant and trust **enable long-term cooperation**, moral resilience, and ethical continuity. Civilizations rise when covenants are sacred and trust is strong; they fall when promises are broken and trust erodes.

**Covenant and Trust: Christian Perspective**

Covenant theology is central to the Bible:

- Noahic Covenant: promise to creation (Genesis 9)

- Abrahamic Covenant: call to faith and blessing (Genesis 12, 15)

- Mosaic Covenant: moral law and communal identity (Exodus 20)

- New Covenant in Christ: fulfillment through grace and internal transformation (Luke 22:20)

God is portrayed as faithful to covenant, even when humanity is not.

Trust is expressed in:

- Faith in God's promises

- Relational loyalty

- Community formed through sacrificial love

Paul writes:

> *"It is required that those who have been given a trust must prove faithful."* (1 Corinthians 4:2)

Christian covenant is both vertical (with God) and horizontal (with others)—a sacred web of relational fidelity.

## Covenant and Trust: Islamic Perspective

Islam is structured around a primordial covenant between God and humanity:

> *"Am I not your Lord?" They said, "Yes, we bear witness."* (Qur'an "The Heights" 7:172)

This original covenant is reaffirmed through:

- Prophethood and revelation

- Shahādah (the declaration of faith)

- Contracts in marriage, business, and governance

Trust (*amānah*) is a divine trust placed upon humans:

*"We offered the Trust to the heavens and the earth... but man undertook it. Indeed, he was unjust and ignorant."* (Qur'an "The Confederates" 33:72)

The Prophet Muhammad said:

*"The signs of a hypocrite are three... when he is entrusted, he betrays."* (Bukhari & Muslim)

Islamic law enshrines:

- Contractual justice

- Mutual rights and responsibilities

- God as the ultimate Witness and Judge of trust

| Criterion | Assessment |
|---|---|
| **1. Communication** | ✔ Creates moral transparency and shared narrative. |
| **2. Friction** | ✔ Reduces suspicion, betrayal, and institutional failure. |
| **3. Problem Solving** | ✔ Supports long-term trust needed for ethical governance and repair. |
| **4. Resilience** | ✔ Fosters durability in personal, communal, and spiritual relationships. |

| | |
|---|---|
| **5. Trust & Cooperation** | ✓ Covenant forms the moral basis for fidelity, empathy, and shared vision. |
| **6. Adaptability** | ✓ Covenant frameworks can evolve while retaining their ethical essence. |
| **7. Pro-Social w/o Reward** | ✓ Faithfulness is often practiced at cost, out of moral duty. |
| **8. Functional Health** | ✓ Healthy covenants create stable institutions and loyal relationships. |

**Covenant and Trust: Does it Align with Future Human Growth and Evolution?**

**Covenant and Trust: Points of Alignment**

- Both Islam and Christianity treat covenant and trust as **sacred realities**, not social contracts alone.

- Covenant is the spiritual form of commitment—a bond of love, law, and moral clarity.

- Trust is what allows human freedom to be coordinated, not exploited.

**Covenant and Trust: Points of Divergence or Nuance**

- Islam grounds covenant in law, accountability, and mutual duty, with trust enforced through divine oversight and communal ethics.

- Christianity views covenant as grace-based and fulfilled in Christ, stressing relational transformation and trust rooted in divine fidelity.

- Both traditions warn against betrayal of trust—not merely as error, but as a wound to the sacred order.

**Covenant and Trust Conclusion: Evolutionary Trajectory**

Covenant and trust are the moral architecture of evolved life. Without them, intimacy collapses into suspicion, leadership degrades into manipulation, and societies fragment into disillusionment and self-interest. Islam and Christianity both call humanity back to a higher vision, where promises are not expedient but sacred, and trust is not naïve but a courageous act of ethical faith. In the evolutionary-spiritual framework, covenant is not a relic of ancient religion; it is a living instrument of moral coherence, essential for sustainable relationships, just governance, resilient communities, and inner integrity. Covenant binds us not through coercion, but through shared commitment to dignity, accountability, and transcendent purpose. In the future shaped by ethical evolution, **trust will be seen as the oxygen of civilization**, fragile yet foundational, and covenant as the spiritual scaffolding that **holds together not only beliefs but entire cultures**. It is through these enduring bonds that humanity will move **from fear to fidelity**, and from fragmentation to wholeness.

# Chapter 40: Prayer

**Prayer: Definition and Purpose**

Prayer is the intentional act of communicating with the Divine—through praise, petition, confession, reflection, or silent presence. It is a discipline of reverence, alignment, and surrender that connects the human soul to its source, purpose, and moral compass.

In the evolutionary-spiritual framework, prayer is both a personal anchor and a communal rhythm. It fosters emotional regulation, ethical clarity, humility, and spiritual attentiveness. Through prayer, the human being is reminded that power is not ultimate, presence is sacred, and silence can be holy.

**Prayer: Christian Perspective**

Prayer in Christianity is intimate, relational, and transformative. Jesus frequently prays—alone, with others, and in times of joy, sorrow, and struggle.

> *"When you pray, say: 'Our Father in heaven…'"* (Luke 11:2)

The Lord's Prayer models:

- Reverence: "Hallowed be your name"

- Submission: "Your will be done"

- Dependence: "Give us this day our daily bread"

- Forgiveness and moral accountability

Christian prayer includes:

- Spontaneous petitions and intercessions

- Contemplative silence and listening

- Liturgy and sacraments (e.g., the Eucharist as prayer in action)

Paul teaches:

> *"Pray without ceasing."* (1 Thessalonians 5:17)

Prayer is framed as a relationship, not a ritual—a way to dwell in God's presence.

## Prayer: Islamic Perspective

Prayer (*ṣalāh*) is the second pillar of Islam—an obligatory act performed five times daily as a direct link between the believer and God.

> *"Establish prayer to remember Me."* (Qur'an 20:14)

Distinct forms of prayer include:

- Ṣalāh: formal, structured, physical-spiritual ritual performed at set times

- Du'ā': personal supplication, spontaneous and heartfelt

- Dhikr: remembrance of God, often repeated phrases glorifying the Divine

The Prophet Muhammad said:

> *"Prayer is the pillar of religion."* (Bayhaqī)

Prayer in Islam is:

- A spiritual discipline: teaches humility, focus, and patience

- A communal bond: Friday prayers, Eid, and congregational ṣalāh foster unity

- A moral recalibration: regular intervals of remembrance reduce heedlessness, anchor gratitude, and prevent sin

| Criterion | Assessment |
|---|---|
| **1. Communication** | ✔ Deepens connection with the Divine and enhances inner moral dialogue. |
| **2. Friction** | ✔ Cultivates peace, patience, and relational empathy. |
| **3. Problem Solving** | ✔ Enhances reflection, intention, and ethical clarity. |
| **4. Resilience** | ✔ Anchors spiritual strength in hardship and gratitude in abundance. |
| **5. Trust & Cooperation** | ✔ Strengthens moral communities and shared sacred rituals. |
| **6. Adaptability** | ✔ Forms and expressions of prayer evolve across time and culture. |
| **7. Pro-Social w/o Reward** | ✔ Rooted in sincerity, not visibility or reward. |

| | |
|---|---|
| **8. Functional Health** | ✔ Associated with lower stress, better self-regulation, and increased empathy. |

## Prayer: Does it Align with Future Human Growth and Evolution?

### Prayer: Points of Alignment

- Both Islam and Christianity view prayer as central to spiritual life, not optional.

- Prayer is not magic or transaction—it is presence, communion, and transformation.

- In both traditions, prayer forms the ethical heart, aligning daily action with sacred reality.

### Prayer: Points of Divergence or Nuance

- Islam ritualizes prayer in precise form and timing, emphasizing discipline and embodiment.

- Christianity allows for more personal spontaneity, emphasizing intimacy and relational depth.

- Islam emphasizes structured surrender; Christianity emphasizes spiritual dialogue—but both converge on prayer as sacred encounter.

### Prayer: Conclusion: Evolutionary Trajectory

Prayer is how the human spirit orients itself toward the sacred, how civilizations tune themselves to justice, and how consciousness returns to its deepest source. **Islam and Christianity call us to prayer not because God requires affirmation, but because we require alignment, daily inward calibration that keeps ego in check and soul attuned to the**

**eternal.** In the evolutionary-spiritual framework, prayer is more than ritual or refuge; it is an essential rhythm of inner evolution. It **refines attention, nurtures humility, and reconnects the self** to something higher than impulse or ideology. As humanity advances, prayer will be rediscovered not as withdrawal from the world, but as the engine of ethical clarity and sacred intention. It will become the still point at the center of action, the place from which wisdom radiates, compassion is renewed, and the fragmented human story begins to harmonize with a deeper cosmic calling. In this future, to **pray is not to escape, but to evolve, one breath, one pause, one sacred word at a time**.

# Chapter 41: Sacrifice

**Sacrifice: Definition and Purpose**

Sacrifice is the **intentional act of giving up something valuable,** such as comfort, time, power, wealth, or even life, for a higher moral, spiritual, or communal purpose. It is one of the deepest expressions of love, loyalty, and conviction. At its root, sacrifice is not loss; it is offering.

In the evolutionary-spiritual framework, sacrifice is a signal of sincerity, altruism, and belonging. It strengthens moral bonds, cultivates resilience, and reveals the values a person or society holds most sacred. When practiced rightly, sacrifice elevates the soul and stabilizes civilization.

**Sacrifice: Christian Perspective**

Christianity centers on the sacrificial love of Jesus Christ, whose crucifixion is seen as the ultimate act of self-giving for the redemption of humanity.

> *"Greater love has no one than this: to lay down one's life for one's friends."* (John 15:13)

Sacrifice in Christianity includes:

- Spiritual sacrifice: daily offering of self in service and worship (Romans 12:1)

- Moral sacrifice: choosing integrity over convenience, forgiveness over vengeance

- Communal sacrifice: sharing time, resources, and compassion with the suffering

The cross is both:

- A symbol of divine love

- A model for human action—sacrificial love that transforms suffering into salvation

## Sacrifice: Islamic Perspective

Sacrifice is central to Islamic theology and practice. The word *qurbān* (sacrifice) shares its root with *qurb*—closeness to God.

Key expressions include:

- Eid al-Adha: Commemorates Abraham's willingness to sacrifice his son, redirected by God to a ram (Qur'an "Those who set the Ranks" 37:102–107)

- Fasting in Ramadan: Sacrificing food and comfort for spiritual purification

- Charity and zakāt: Sacrificing wealth to lift others

*"Never will you attain righteousness until you spend from that which you love."* (Qur'an "The Family of Imrān" 3:92)

Islamic sacrifice is:

- **Voluntary** yet deeply commanded

- Purifying: removing ego and attachment

- **Community-oriented**: building solidarity through shared giving

The Prophet Muhammad emphasized that the greatest sacrifices are those done in secret, for the sake of God alone.

| Criterion | Assessment |
|---|---|
| 1. Communication | ✔ Demonstrates sincerity, deep values, and ethical intent. |
| 2. Friction | ✔ Subverts competition by modeling generosity and self-restraint. |
| 3. Problem Solving | ✔ Enables moral prioritization and **systemic commitment.** |
| 4. Resilience | ✔ Builds endurance through discipline and shared hardship. |
| 5. Trust & Cooperation | ✔ Signals loyalty and creates strong communal bonds. |
| 6. Adaptability | ✔ Flexible in form but stable in moral significance. |
| 7. Pro-Social w/o Reward | ✔ Often practiced at personal cost with no immediate return. |
| 8. Functional Health | ✔ Reduces selfishness and promotes interdependence. |

**Sacrifice: Does it Align with Future Human Growth and Evolution?**

**Sacrifice: Points of Alignment**

- Both Islam and Christianity teach that sacrifice is the path to spiritual maturity.

- **Sacrifice is not about destruction.** It is about sanctification, revealing what we love by what we're willing to give.

- Sacrificial action is considered the truest test of sincerity, whether toward God or neighbor.

**Sacrifice: Points of Divergence or Nuance**

- Islam views sacrifice as ethical discipline and ritual devotion, with practices embedded in daily life, festivals, and charity.

- Christianity emphasizes theological sacrifice—rooted in grace and redemption through Christ's crucifixion, then mirrored in a life of sacrificial love.

- Islam encourages communal sacrifice as a test of gratitude; Christianity frames it as participation in divine suffering and love.

**Sacrifice Conclusion: Evolutionary Trajectory**

Within the evolutionary-spiritual framework, sacrifice serves as an essential catalyst for ethical advancement, marking the transition from biological impulse toward spiritual maturity and moral refinement. It demonstrates that human evolution involves not merely survival or self-interest, but a conscious commitment to higher purposes and collective well-being. Both Islam and Christianity strongly support this understanding by highlighting sacrifice as transformative, not diminishing individuals but elevating their spiritual and ethical capacities. Sacrifice, in this context, embodies sincerity by turning abstract beliefs into tangible actions that enhance communal resilience, deepen trust, and fortify cooperative bonds. **This willingness to offer something valuable for the greater good reflects humanity's potential to transcend self-centered instincts and foster deeper relational integrity and social cohesion**. As humanity advances on its evolutionary path, sacrifice will increasingly be viewed as a powerful evolutionary virtue, a sacred strength that transforms instinctual desires into sustained moral action, and individual aspirations into collective flourishing.

# Epilogue

## Toward a Spiritually Evolved Humanity

This journey began with a question: If humans are capable of exerting influence on their own future evolutionary development, can the moral and spiritual teachings of most religions, like those found in Christianity and Islam, offer not only meaning but a roadmap for human evolution? Whether coming from a scientific or religious background, can we appreciate the moral lessons presented in these religious traditions **without changing our existing belief structure**? Would these lessons be worthwhile to incorporate into our thoughts and actions for the purpose of our **collective future growth**? We have examined more than thirty ethical principles, drawn from scripture, tradition, and social memory. We have evaluated them not only in theological terms but also in their capacity to support resilience, cooperation, emotional intelligence, and communal flourishing. We have asked not only whether they are right, but whether they help us grow.

Across these chapters, a pattern has emerged. The virtues most emphasized in sacred texts, such as justice, mercy, patience, love, humility, and truth, are also those that **enhance the long-term survivability of societies**. The moral structure of revelations align closely with the demands of an interconnected, fragile, and evolving human ecosystem. What religion calls righteousness, evolution may call sustainability. What theology calls grace, neuroscience may call emotional regulation. What scripture calls sin; sociology may call social entropy. Different languages, same underlying truths.

This book has not sought to demonstrate that religion is scientifically correct or that science can validate theology. Rather, it has proposed that the two are not enemies but **partners**, each illuminating different aspects of the human condition. Religion, when interpreted through its ethical core, is not opposed to progress; it is progress, encoded in moral form. And science, when guided by conscience, is not spiritually hollow; it is spiritually potent.

The evolutionary-spiritual model proposed here is not a theory of biology. It is a vision of humanity's ethical development, one that recognizes that our next great leap will not come from technology alone, but from the choices we make about how we live, what we value, and who we become.

The future belongs to communities that can cooperate across differences, tell the truth in love, care for the vulnerable, forgive one another, seek justice, and cultivate inner peace. It belongs to those who can integrate ancient wisdom with contemporary insight. It belongs to those who can say, **"We were made for more."**

This model does not demand agreement on metaphysical doctrine. It invites convergence on moral direction. And it challenges each of us, believer, skeptic, seeker, and scholar alike, to examine how our actions shape the long arc of human development.

**The Work Ahead**

The task now is not merely to demonstrate and admire the alignment between revelation and evolution; it is to embody it. To build families, institutions, schools, and societies that reflect this synthesis of ethics and adaptability. To recognize that spiritual growth is not a private pursuit, but a public necessity. And to understand that every act of kindness, every truth spoken with love, every injustice challenged with courage, **these are evolutionary events**. These are the true revolutions.

In this way, the sacred and the scientific meet, not in contradiction but in harmony. Not in abstract theory, but in the moral architecture of daily life.

May we walk that path with wisdom.

May we build communities that endure.

And may we, through love, truth, and mercy, become the kind of people our ancestors prayed for, and **our descendants will remember with gratitude.**

# Appendices

# Appendix 1

**8-Point Empirical Rubric for Evaluating Religious or Philosophical Principles against the Evolutionary-Spiritual Framework.**

1. **Improves <u>communication</u> and mutual understanding**

   > Does it enhance the clarity, transparency, and frequency of communication between individuals and groups, especially across cultures?

2. **Reduces interpersonal and societal <u>friction</u>**

   > Does it lead to measurable reductions in conflict, hostility, mistrust, or polarization between individuals and social groups?

3. **Enhances collective <u>problem-solving</u> capacity**

   > Does it increase the ability to identify, analyze, and solve problems between individuals or groups?

4. **Strengthens long-term stability and adaptive <u>resilience</u>**

   > Does it help individuals and societies endure stress, recover from disruption, and maintain structural or relational continuity over time?

5. **Fosters affinity, <u>trust</u>, and cooperation across boundaries**

Does it increase the likelihood of voluntary cooperation, mutual aid, or resource-sharing between individuals and groups, even across ethnic, religious, or national lines?

6. **Is scalable and <u>adaptable</u> across cultures and futures**

Can the principle be applied in different cultural, technological, or ecological contexts without degrading its core function or causing harm?

7. **Encourages <u>pro-social behavior</u> without external enforcement**

Does it motivate actions like honesty, generosity, or restraint even when there is no surveillance, punishment, or reward involved?

8. **Improves <u>functional health</u> and reduces signs of social dysfunction**

Does it lead to reductions in measurable stress indicators (e.g., violence, burnout, disconnection) and increase capacities like participation, cohesion, or initiative?

# Appendix 2

## Comprehensive Glossary of Terms

Adl: Justice; a core value in Islam, encompassing fairness in all matters.

Acetylcholine: A neurotransmitter involved in muscle activation, memory, and cognition.

Agapē: Greek for unconditional, divine love in Christian theology.

Āmānah: Moral trust or entrusted responsibility; an ethical obligation.

Asabiyyah: Tribalism or ethnocentric loyalty, discouraged in Islamic ethics.

Autocatalytic: Describes a chemical reaction in which one of the products serves as a catalyst for the reaction itself, often central to early self-replicating systems.

'Afw: Forgiveness or pardon, a higher moral act beyond justice.

Bukhari: One of the most authentic collections of hadith in Sunni Islam.

Caritas: Latin for charity, one of the three theological virtues.

Charter of Medina: A historic constitution drafted by Prophet Muhammad that established rights and duties for Muslims, Jews, and other communities in Medina, fostering pluralism and mutual justice.

Collective Consciousness: A shared set of beliefs, ideas, and moral attitudes that operate as a unifying force within society or a species.

Dajjāl: The Islamic antichrist figure expected before the Day of Judgment.

Darwinian: Relating to Charles Darwin's theories of natural selection and biological evolution based on survival and reproductive fitness.

Dhikr: Remembrance of God through repeated phrases or meditative invocation.

Dhiyāfah: Hospitality; the act of generously welcoming guests.

Du"ā: Personal supplication or prayer in Islam.

ENS: Enteric Nervous System; a complex network of neurons embedded in the gastrointestinal tract, often referred to as the body's 'second brain'.

Endosymbiosis: A symbiotic relationship where one organism lives inside another; key to the origin of eukaryotic cells.

Entropy: A measure of disorder or randomness in a system; in thermodynamics, systems evolve towards greater entropy.

Eucharistia: Greek for thanksgiving, the source word for the Eucharist in Christianity.

Eukaryotes: Organisms whose cells contain a nucleus and organelles enclosed within membranes.

Evolutionary Nudge: A metaphorical term describing moral or religious guidance that promotes behaviors enhancing long-term survival and evolution.

GABA: Gamma-Aminobutyric Acid; a neurotransmitter in the central nervous system that inhibits nerve transmission and helps regulate anxiety and relaxation.

Ghaḍab: Anger; considered a dangerous passion to be restrained.

Gibbs: Gibbs Free Energy; a thermodynamic function predicting whether a system's process will occur spontaneously.

Group Selection: An evolutionary theory proposing that natural selection can act on groups, favoring traits that benefit the group, even if costly to individuals.

Hadith: Recorded sayings, actions, and approvals of the Prophet Muhammad, second in authority only to the Qur'an in Islamic tradition.

Ḥalāl: Permissible or lawful according to Islamic law.

Ḥarām: Forbidden or prohibited by Islamic law.

Hasad: Envy; a morally condemned inner disease in Islam.

Ikhlāṣ: Sincerity or purity of intention, essential for accepted worship.

Immortality Project: A term from Ernest Becker's work, describing human efforts to achieve symbolic immortality through legacy, belief, or culture.

Iqra: The first word revealed in the Qur'an, meaning 'Read'.

Isomers: Molecules with the same chemical formula but different structures and properties.

Isrāf: Extravagance; overindulgence in consumption, discouraged in Islam.

Jahannam: Hell in Islamic theology.

Jannah: Paradise in Islamic belief.

Jihād al-Nafs: The internal struggle against the self's lower desires.

Jumuʿah: Friday congregational prayer, a sacred time in Islam.

Karām: Generosity; a moral and spiritual trait highly praised in Islam.

Kasl: Laziness or spiritual lethargy.

Khalifa: An Islamic term meaning steward or vicegerent; used in the Qur'an to describe humanity's role as caretaker of Earth.

Khums: A Shia obligation to give one-fifth of surplus wealth annually.

Kibr: Arrogance; a sin that led to Iblīs 'fall.

Kin Selection: An evolutionary mechanism favoring behaviors that increase the reproductive success of relatives.

Kitāb: The divine record of each person's deeds.

Mahdī: Prophesied redeemer in Islamic eschatology.

Marājiʿ: High-ranking religious authorities in Shia Islam.

Metabolites: Small molecules involved in metabolism, often produced or influenced by microbiota, with effects on health and behavior.

Microbiome: The collective genetic material of microorganisms living in a particular environment, such as the gut or skin.

Mīzān: The scale used to weigh deeds on the Day of Judgment.

Murābaḥah: Islamic cost-plus-profit sale contract.

Mushārakah: Profit-sharing business partnership in Islamic finance.

Natural selection: A Darwinian concept where traits that improve survival and reproduction become more common in a population.

Neurotransmitters: Chemicals that transmit signals across a synapse from one neuron to another in the nervous system.

Nifāq: Hypocrisy; considered a grave spiritual failing.

Niyyah: Intention; determines the spiritual value of acts in Islam.

Nostra Aetate: Latin document from Vatican II promoting interfaith respect.

Prokaryotes: Single-celled organisms without a nucleus, such as bacteria and archaea.

Qanāʿah: Contentment and inner sufficiency.

Qiyāmah: The Day of Resurrection and final judgment in Islam.

Quran: The holy book of Islam, believed by Muslims to be the literal word of God as revealed to Prophet Muhammad.

Raj'ā: Hope in God's mercy and providence.

Reciprocal Altruism: A behavior whereby an organism acts in a way that temporarily reduces its fitness while increasing another's, with the expectation of future reciprocation.

Ribā: Usury; prohibited gain on loans in Islamic finance.

Riy'ā: Showing off in religious acts; a form of hidden shirk.

Ṣabr: Patience and endurance in the face of adversity.

Ṣadaqah: Voluntary charity given beyond the obligatory zakāt.

Serotonin: A neurotransmitter that regulates mood, sleep, and appetite; largely produced in the gastrointestinal system.

Shahwah: Desire, especially sexual or physical craving.

Sharīʿah: Islamic legal framework guiding both public and private life.

Shuʿaib: A prophet in Islam known for advocating economic justice and honesty in trade, mentioned in the Qur'an.

Shūrā: Mutual consultation, particularly in leadership and governance.

Ṣidq: Truthfulness; a mark of integrity and moral excellence.

Sociobiology: A field of biology that explores the biological basis of social behavior across species, including humans.

Spiritual Prescriptions: Religious or moral directives that encourage behaviors believed to benefit individuals and society both spiritually and practically.

Ṣulḥ: Reconciliation; making peace between conflicting parties.

Ṣulḥ al-Ḥudaybiyyah: A pivotal treaty between Muslims and Quraysh in early Islam.

Symbiosis: A close and often long-term interaction between two different biological species.

Taqdīr: Divine decree or destiny.

Taqwā: God-consciousness; moral vigilance and spiritual awareness.

Tawāḍuʿ: Humility and self-effacement.

Tawakkul: Reliance on God after taking appropriate effort.

Tazkiyah: Spiritual purification of the heart.

Thermodynamics: A branch of physics dealing with heat, work, and energy transformations.

Ummah: The global Muslim community.

Ummatan Wasaṭan: A balanced or moderate community (Qur'an 2:143).

Wasatiyyah: Moderation and balance in all aspects of life.

Yatāmā: Orphans; protected and honored in Islamic ethics.

Zakat: A mandatory form of almsgiving in Islam, one of the Five Pillars, aimed at redistributing wealth and supporting the community.

# Appendix 3

**Religion References**

The following references include scriptural citations and classical sources cited throughout the manuscript. They are formatted for inclusion in the final publication bibliography.

**The Holy Bible (Old Testament)**

- **Genesis 1:31**

- **Genesis 2:15**

- **Deuteronomy 6:7**

- **Proverbs 9:10**

- **Proverbs 12:10**

- **Isaiah 1:17**

- **Isaiah 53:5**

- **Amos 5:24**

**The Holy Bible (New Testament)**

- **Matthew 4:17**

- **Matthew 5:3**

- Matthew 5:7

- Matthew 5:9

- Matthew 5:37

- Matthew 5:44

- Matthew 6:5

- Matthew 6:12

- Matthew 6:24

- Matthew 15:8

- Matthew 16:24

- Matthew 19:5

- Matthew 22:37

- Matthew 22:39

- Matthew 23:11

- Matthew 23:12

- Matthew 25:13

- Matthew 25:35

- Matthew 25:40

- Matthew 26:52

- 1 Corinthians 4:2

- 1 Corinthians 6:12

- 1 Corinthians 6:19

- 1 Corinthians 13:3

- 1 Corinthians 13:13

- 2 Corinthians 1:3

- 2 Corinthians 5:18

- Galatians 5:22

- Galatians 6:9

- Ephesians 4:25

- Philippians 2:8

- Philippians 3:10

- Philippians 3:20

- 1 Timothy 6:6

- Hebrews 12:1

- Hebrews 13:2

- 1 Peter 5:5

**The Qur'an**

- Al-Baqarah (The Cow) 2:143

- Al-Baqarah (The Cow) 2:153

- Al-Baqarah (The Cow) 2:190

- Al-Baqarah (The Cow) 2:214

- Al-Baqarah (The Cow) 2:256

- Al-Baqarah (The Cow) 2:275

- An-Nisa (The Women) 4:58

- Al-Ma'idah (The Table Spread) 5:90

- Al-An'am (The Cattle/Livestock) 6:38

- Al-An'am (The Cattle/Livestock) 6:165

- Al-Anfal (The Bounties) 8:61

- Ibrahim (Abraham) 14:7

- An-Nahl (The Bee) 16:90

- An-Nahl (The Bee) 16:125

- Maryam (Mary) 19:96

- Ta-Ha 20:14

- Ta-Ha 20:131

- Al-Ahzab (The Confederates) 33:72

- Az-Zumar (The Crowds) 39:9

- Az-Zumar (The Crowds) 39:53

- Ash-Shura (The Consultation) 42:40

- Muhammad 47:19

- Al-Hujurat (The Chambers) 49:13

- Al-Mumtahanah (She that is to be Examined) 60:8

- Al-Jumu'ah (Friday) 62:9

- Al-Bayyinah (The Clear Proof) 98:5

- Al-Ma'un (Small Kindnesses) 107:1

## Hadith Collections

- Abu Dawud

- Bayhaqi

- Bukhari

- Bukhari & Muslim (Muttafaqun ʿalayhi)

- Ibn Mājah

- Muslim (Ṣaḥīḥ Muslim)

- Nasāʾī

- Ali Ibn Abi Talib (Nahjul Balagha)

- **Tirmidhi**

**Notes:**

- **Bible translations:**
  Unless otherwise stated, biblical quotes are from a common, contemporary translation (e.g., NIV, ESV, KJV).

- **Qur'anic translations:**
  Unless otherwise specified, quotes from the Qur'an are from a widely accepted translation such as Sahih International, Yusuf Ali, or Pickthall.

# Appendix 4

References for Chapter 2: The Science – Evolution from Cosmological Origins to Human Behavior

1. Cosmological Origins and Particle Physics

- Greene, B. (2004). *The Fabric of the Cosmos: Space, Time, and the Texture of Reality.* New York: Knopf.

- Carroll, S. (2010). *From Eternity to Here: The Quest for the Ultimate Theory of Time.* New York: Dutton.

- Kolb, E. W., & Turner, M. S. (1994). *The Early Universe.* Boulder, CO: Westview Press.

2. Chemical Evolution and Molecular Complexity

- Hazen, R. M. (2005). *Genesis: The Scientific Quest for Life's Origins.* Washington, DC: Joseph Henry Press.

- Miller, S. L., & Urey, H. C. (1959). *Organic compound synthesis on the primitive Earth.* Science, 130(3370), 245–251.

- Fry, I. (2000). *The Emergence of Life on Earth: A Historical and Scientific Overview.* Rutgers University Press.

3. Thermodynamics and Self-Organization

- Prigogine, I., & Stengers, I. (1984). *Order Out of Chaos: Man's New Dialogue with Nature.* New York: Bantam Books.

- England, J. L. (2013). *Statistical physics of self-replication.* The Journal of Chemical Physics, 139(12), 121923.

4. Biological Evolution and Complexity

- Dawkins, R. (2009). *The Greatest Show on Earth: The Evidence for Evolution.* New York: Free Press.

- Margulis, L. (1998). *Symbiotic Planet: A New Look at Evolution.* New York: Basic Books.

- Lane, N. (2015). *The Vital Question: Energy, Evolution, and the Origins of Complex Life.* New York: W. W. Norton & Company.

5. Emergence of Human Intelligence and Society

- Tomasello, M. (2014). *A Natural History of Human Thinking.* Harvard University Press.

- Wilson, E. O. (2012). *The Social Conquest of Earth.* New York: Liveright.

6. Epigenetics and Behavioral Inheritance

- Jablonka, E., & Lamb, M. J. (2005). *Evolution in Four Dimensions: Genetic, Epigenetic, Behavioral, and Symbolic Variation in the History of Life.* MIT Press.

- Meaney, M. J. (2001). *Maternal care, gene expression, and the transmission of individual differences in stress reactivity across generations.* Annual Review of Neuroscience, 24(1), 1161–1192.

· Dias, B. G., & Ressler, K. J. (2014). *Parental olfactory experience influences behavior and neural structure in subsequent generations.* Nature Neuroscience, 17(1), 89–96.

## 7. Human Society and Collective Intelligence

· Pentland, A. (2014). *Social Physics: How Social Networks Can Make Us Smarter.* Penguin Press.

· Christakis, N. A., & Fowler, J. H. (2010). *Connected: The Surprising Power of Our Social Networks and How They Shape Our Lives.* New York: Little, Brown and Company.

## 8. Neurobiology of the Gut-Brain Axis

· Mayer, E. A. (2016). *The Mind-Gut Connection: How the Hidden Conversation Within Our Bodies Impacts Our Mood, Our Choices, and Our Overall Health.* New York: Harper Wave.

· Cryan, J. F., & Dinan, T. G. (2012). *Mind-altering microorganisms: the impact of the gut microbiota on brain and behaviour.* Nature Reviews Neuroscience, 13(10), 701–712.

# Appendix 5

Scientific References on the Health Impact of Virtues and Vices

The following peer-reviewed studies and foundational works provide empirical support for the proposition that virtues such as gratitude, love, forgiveness, contentment, and humility offer tangible health benefits— psychological, physiological, and social. Similarly, they illustrate how unchecked vices may correspond with measurable declines in individual and communal well-being.

## Gratitude

- Emmons, R. A., & McCullough, M. E. (2003). *Counting blessings versus burdens: An experimental investigation of gratitude and subjective well-being in daily life.* Journal of Personality and Social Psychology, 84(2), 377–389.

- Wood, A. M., Froh, J. J., & Geraghty, A. W. A. (2010). *Gratitude and well-being: A review and theoretical integration.* Clinical Psychology Review, 30(7), 890–905.

## Love, Connection, and Social Bonds

- Cacioppo, J. T., & Patrick, W. (2008). *Loneliness: Human Nature and the Need for Social Connection.* New York: W. W. Norton & Company.

- Uchino, B. N. (2006). *Social support and health: A review of physiological processes potentially underlying links to disease outcomes.* Journal of Behavioral Medicine, 29(4), 377–387.

## Forgiveness and Reconciliation

- Witvliet, C. V. O., Ludwig, T. E., & Vander Laan, K. L. (2001). *Granting forgiveness or harboring grudges: Implications for emotion, physiology, and health.* Psychological Science, 12(2), 117–123.

## Mindfulness and Contentment

- Kabat-Zinn, J. (1994). *Wherever You Go, There You Are: Mindfulness Meditation in Everyday Life.* New York: Hyperion.

- Brown, K. W., & Ryan, R. M. (2003). *The benefits of being present: Mindfulness and its role in psychological well-being.* Journal of Personality and Social Psychology, 84(4), 822–848.

## Humility and Psychological Resilience

- Tangney, J. P. (2000). *Humility: Theoretical perspectives, empirical findings, and directions for future research.* Journal of Social and Clinical Psychology, 19(1), 70–82.

## The Impact of Vice (Stress, Greed, Addiction)

- Sapolsky, R. M. (2004). *Why Zebras Don't Get Ulcers.* New York: Henry Holt and Company.

- Kassirer, J. P. (1995). *The ethics of managed care.* New England Journal of Medicine, 332, 1326–1328.

# Appendix 6

The Charter of Medina

*(c. 622 CE – Neutral English Translation)*

This document is a formal agreement between various tribal, religious, and ethnic communities in the city of Yathrib, later known as Medina. It outlines a shared framework for coexistence, mutual responsibility, and governance among Muslims, Jews, and other allied groups who had previously lived under separate codes of conduct. The charter recognizes them as a single political entity for matters of collective interest and security.

**Key Articles**

1. **Unified Community**
   The signatories form a single political community, despite religious and ethnic differences. They are expected to support one another in upholding justice, maintaining order, and protecting Medina.

2. **Mutual Protection and Responsibility**
   If any party to the agreement is attacked or wronged, all others are obligated to provide assistance. Protection is extended across tribal and religious lines, based on mutual obligation rather than blood relations.

3. **Religious Autonomy and Equality**
   Each religious group, including Jewish tribes, retains the

freedom to practice its faith and manage its own internal affairs. Political equality is granted so long as all groups adhere to the terms of the collective agreement.

4. **Justice and Accountability**
   Wrongdoing is not protected by tribal affiliation. Justice must be applied impartially, and those who commit injustice bear the consequences individually.

5. **Support for the Vulnerable**
   Members of the community must care for the disadvantaged, including orphans, the poor, and those in debt. No one is to be left unprotected or abandoned.

6. **Economic and Legal Coordination**
   Tribal groups are responsible for fulfilling their legal and financial obligations, such as blood-money payments. These duties are to be managed transparently and in accordance with agreed rules.

7. **Sanctity of Medina**
   Medina is designated as a protected and inviolable territory. Armed conflict and acts of betrayal within its boundaries are prohibited. Safety and order must be maintained by all signatories.

8. **External Relations**
   No individual group may establish separate treaties or alliances with outside forces without collective consultation. Any decision involving conflict or peace beyond Medina must be jointly agreed upon.

9. **Dispute Resolution**
   When serious disagreements arise that cannot be resolved through customary means, the matter is to be referred to a central authority recognized by the community. At the time of

the charter's drafting, this role was assigned to Muhammad ibn Abdullah as a neutral arbiter.

10. **Commitment to Peace and Cooperation**
    All signatories are expected to refrain from supporting injustice, betrayal, or aggression. Anyone who violates the agreement may lose the protections it provides. Maintaining the peace and well-being of the broader community is a shared obligation.